Introduction to
Transportation Systems

For a listing of recent titles in the *Artech House ITS Library*,
please turn to the back of this book.

Introduction to Transportation Systems

Joseph Sussman

Artech House
Boston • London
www.artechhouse.com

Library of Congress Cataloging-in-Publication Data
Sussman, Joseph.
 Introduction to transportation systems / Joseph Sussman.
  p.   cm. — (Artech House ITS library)
 Includes bibliographical references and index.
 ISBN 1-58053-141-5 (alk. paper)
 1. Transportation.   I. Title.   II. Series.
 HE151.S775   2000
 388—dc21   00-027451
                            CIP

British Library Cataloguing in Publication Data
Sussman, Joseph
 Introduction to transportation systems. — (Artech House
 intelligent transportation systems library)
 1. Transportation   2. Transportation engineering   3. Intelligent
 Vehicle Highway Systems
 I. Title
 388.3
 ISBN 1-58053-141-5

Cover design by Igor Valdman. Text design by Darrell Judd.
Photo of Dr. Joseph Sussman courtesy of Donna Coveney.
Photos used on pages 1, 147, and 277 courtesy of Dr. Joseph Sussman.

© 2000 Artech House, Inc.
685 Canton Street
Norwood, MA 02062

International Standard Book Number: 1-58053-141-5
Library of Congress Catalog Card Number: 00-027451

10 9 8 7 6 5 4 3 2 1

To Henri-Ann, my love and my partner in life.

Contents

PART II: Freight Transportation   147

12   The Logistics System and Freight Level-Of-Service   149

13   Railroads: Introductory Concepts   171

23 Traveler Level-of-Service   307

24 Intelligent Transportation Systems (ITS)   317

25   The Urban Transportation Planning Process and Real-Time Network Control

Preface

> *… (t)he obvious is precisely what needs to be pointed out—otherwise, it will be overlooked.*
>
> —Peter F. Drucker

In 1994, as a member of the MIT transportation faculty, then for some 26 years, I had previously had several opportunities during the 1970s and 1980s to teach the introductory subject in MIT's graduate transportation program. In that year I was once again tapped to teach this subject, which is required for all first-year graduate students in the MST (Master of Science in Transportation) program. But this time, it was different.

I had recently had the honor (and the work) of serving as chairman of the Executive Committee of the Transportation Research Board (TRB), a unit of the National Research Council. In that capacity I had the opportunity to participate in a number of stimulating discussions on transportation and its role in the world, including various technological, systems, and institutional issues. Every three years TRB produces a "Critical Issues in Transportation" statement and, serendipitously, 1994 was the year. I focused much of my attention on the development of those issues, helping to develop the statement so that it reflected a balanced and foresighted view of the transportation world.

It had been two years since I returned from sabbatical in Washington, where in 1991–1992 I served as one of the writers of the *IVHS America Strategic Plan*, a 20-year plan for research, development, testing, and deployment of what was then Intelligent Vehicle Highway Systems and is now Intelligent Transportation Systems (ITS). I digested that experience, which dealt with how to create a large-scale transportation research and deployment program, based on modern technologies.

In 1991 I had the good fortune to be named the first JR East Professor at MIT (endowed by the East Japan Railway Company). Building a relationship between JR East and MIT gave me new insight into a substantially different transportation world. A long-term assignment concerning traffic congestion in Bangkok also broadened my view.

Against this backdrop I began to teach the subject. Educated by the above experiences to a broader understanding of the subtlety of transportation issues, I took the opportunity to redesign this introductory subject to reflect a broader, more conceptual view of the world of transportation than had been the case in my earlier attempts.

I decided to audiotape the lectures, about 44 hours' worth, and produced a transcript amounting to about 200,000 words. The initial transcript captured the subject in the raw. It is a humbling event for a veteran professor to see in black-and-white all the circumlocutions, backtracking, and lack of clarity in the unexpurgated text. Nonetheless some of the text seemed to carry a clear message about transportation.

In editing the transcript, I felt it was of some value to retain the ebb and flow—the fabric—of the lectures, and their informal, conversational tone. I wanted to try to capture the subject in the large to give the reader a sense of how its intellectual content had developed. The reader will see vestiges of the taped lectures in the students' questions inserted in the text.

From 1995-1999 I taught the subject each fall, and improvements were integrated into the materials. I also had a number of wholly new lectures that I "spliced in."

What this text then represents is one semester's teaching at the introductory graduate level in transportation systems. It describes what I think are the fundamental conceptual elements of a transportation system, and how one begins to go about analyzing and designing particular transportation systems.

This book is composed of three parts: Part I, entitled "Context, Concepts, and Characterization," presents basic material about transportation systems, the context within which they operate, and a characterization of their behavior. Part II deals with freight transportation, and Part III deals with traveler transportation. While Part I lays out the basics of the world of transportation systems, the reader can reasonably skip directly to the material on freight and traveler transportation.

There are some sections that are mathematical in content; the reader need not understand the details of those sections in order to understand subsequent material. At the same time, the reader should note that those mathematical treatments are often not rigorous but rather illustrative. Those inclined toward a deep mathematical approach should pursue the many other available references in the transportation literature.

I hope that students, faculty, and practitioners of the transportation field will find the book of value in their study of and practice in this stimulating area.

Joseph M. Sussman
Cambridge, Massachusetts
April 2000

Acknowledgments

This book builds on more than 30 years of experience in teaching, research, and consulting in the transportation field. During that period, I have had the good fortune to serve with some of the great names—too numerous to mention—in the current transportation world. I would particularly like to thank my faculty colleagues at MIT, as well as professional organizations such as the Transportation Research Board and ITS America. Through these relationships I have seen my own perspectives on the field change from a narrow modeling point of view to a broader-based technology/systems/institutions perspective, which is represented in this book.

Many MIT graduate students helped in the preparation of this manuscript: Camille Tsao, Louisa Yue, Michael Finch, and James Grube participated in typing, formatting, and editing text and diagrams that sometimes became unwieldy. I have a special debt to the teaching assistants who participated in the instruction of the introductory graduate subject, from which this book is derived. That includes, in chronological order beginning in 1994: William Cowart, Daniel Rodríguez, Lisa Klein, Sarah Bush, Chris Conklin, and Elton Lin. Each made their own special contribution to the teaching of the subject and the related modifications to this text. The students who took this subject over these years also played an important role in helping shape my arguments and presentation.

I owe a debt of gratitude to Jan Austin Scott, who has lived through innumerable drafts of this material and has provided a keen eye and a good grammatical sense, as well as good taste, in altering the text. Her help and support during the completion of this book played no small role in what you see before you. Denise Brehm and Terri Lehane also provided valuable assistance in the preparation of early drafts.

While this manuscript has benefited from many people's perusal, editing, commentary, etc., any errors or lack of clarity is surely the fault of the author.

PART I

Context, Concepts, and Characterization

CHAPTER 1

Introduction: Context, Concepts, and Characterization

Introduction

The purpose of this book is to introduce the reader to the field of transportation systems.

We will focus on basic ideas: characterizing transportation systems, defining transportation system components, and introducing the idea of transportation networks. Through a simple example, we will establish thirty transportation key points. Arguably, understanding these thirty key points is essential to a broad grasp of transportation system performance. We introduce the idea of models and frameworks as useful abstractions of transportation systems that we can use to perform transportation system analysis and design. Then, we continue with parts on freight transportation and traveler transportation.

Focus on Basic Principles

The approach here is general and is relevant to all transportation modes and geographic contexts. We present a basic framework for thinking about transportation problem solving.

Of course, there will be ideas that may be interesting to people with particular transportation perspectives, while others may not find some of these of great interest. Our basic premise, however, is that there are some underlying principles which are fundamental to the field of transportation. We will introduce these basic principles.

Transportation: A Broad Field

Transportation is a broad and ubiquitous field—important in a political, social, and economic sense. Everywhere you look, transportation is there. We commute. We receive goods by UPS. We drive to shop at the supermarket—and the goods arrived there by truck. We are impacted by the environmental consequences of transportation operations. Transportation is an integral part of our everyday lives.

Transportation and the Social-Political-Economic Context

Socially, politically, and economically, transportation is important throughout the world. It can be a major public policy lever. The public sector often makes important public policy decisions through transportation investments.

Further, the private sector makes major investments in transportation. For example, many countries (particularly the United States) have built their industrial base largely around the automobile. Automobile manufacturing is one of the fundamental underpinnings of the United States economy.

Transportation is a major employer of people in virtually every country in the world. The largest employer in India is the Indian Railroad. When we think about the efficiency of the Indian railroads (which do not have a great productivity record) and we think about the work force (because clearly we could get the job done with fewer people), we must recognize that from a broader perspective, we are driving a very important source of jobs. In making an optimal decision, it may not be so easy as simply saying, "Let's shrink the railroad workforce."

In the United States, about 13% of the gross domestic product (GDP) is transportation-related. Twenty years ago, that figure was about 20%. The percentage of GDP is slowly falling as more of this country's economy becomes service-related (rather than manufacturing or agriculture). In countries that are less developed, the percentage of GDP related to transportation may be larger.

In transportation, we make long-term infrastructure decisions. We build facilities: railroads and highways like the Central Artery/Tunnel in Boston. These facilities are expensive and long lasting. They have a tremendous impact on all of us.

Transportation has impacts on the people who actually use the transportation system. It has impact on the people that are employed, operate, or own the system. However, it has an impact on people who don't use, work for, or own the system. For example, people who live next to a railroad line may benefit little from it, but are heavily impacted by trains going by every 10 minutes. As another example, the urban population is affected by air quality degradation caused by motor vehicles.

Understanding the Dimensions of Transportation

Transportation is multidimensional. It is useful to think about transportation from the perspectives of:

- Technology;
- Systems;
- Institutions.

Technology

By *technology,* we will refer to propulsion, fuels, guideways (e.g., highways, railroads), and guidance and control systems. Also included are materials and how they are used in the development of both guideways and vehicles.

Systems

The second area is *systems.* Here we will focus on the notion of network analysis and modeling how supply and demand interact to produce flows on transportation networks. We will study the microeconomics that drives the behavior of the transportation system, from both the operator and customer points of view.

Institutions

The third dimension is *institutions.* We will think about institutions as a pragmatic approach to how people get things done in the real world of transportation. Deployment and operation of transportation systems does not happen all by itself; it is done by organizations working within a complex social, political, and economic environment.

Certainly, the Central Artery/Tunnel Project in Boston is not happening all by itself. Government agencies are involved. As transportation professionals, we have to deal with the interaction between, for example, the federal and state levels of government. We have to be concerned with the relationship between the Commonwealth of Massachusetts and the City of Boston, and with the private construction company managing the project. We will also have to be concerned with Boston employers and merchants who want to be sure that the city does not shut down for a decade while construction proceeds.

In other situations, we have to think about the interaction of private profit-making organizations, like railroads and airlines, with government organizations that are responsible for public safety and proper economic protection of the captive customer.

These are institutional questions and considerations. It would be naïve to think that simply because we have the right technology and the right transportation system, it will simply happen. Institutions make it happen.

It's been said that

Transportation systems is not rocket science—it is a lot harder.

This results from the extensive interaction of transportation with broader societal issues. Transportation has economic, environmental, and political implications. As we face an increasingly global economy, fundamental organizational and political changes and the advent of increasingly advanced technologies, transportation will become an even more vital field. It is critical that we understand transportation at a fundamental level. It is in this spirit that we will approach the basics of transportation systems.

Complex, Large, Integrated, Open Systems (CLIOS)

Transportation is an example of a broader class of systems we call CLIOS—complex, large, integrated open systems—with the component terms used as described below [1].

A system is complex when it is composed of a group of related units (subsystems), for which the degree and nature of the relationships are imperfectly known. Its overall emergent behavior is difficult to predict, even when subsystem behavior is readily predictable. The time-scales of various subsystems may be very different (as we can see in transportation—land-use changes, for example, versus operating decisions). Behavior in the long-term and short-term may be markedly different and small changes in inputs or parameters may produce large changes in behavior.

CLIOS have impacts that are large in magnitude, and often long-lived and of large geographical extent.

Subsystems within CLIOS are integrated, closely coupled through feedback loops.

By "open" we mean that CLIOS explicitly include social, political, and economic aspects.

Often CLIOS are counterintuitive in their behavior. At the least, developing models that will predict their performance can be very difficult to do. Often the performance measures for CLIOS are difficult to define and, perhaps, even difficult to agree about, depending upon your viewpoint. In CLIOS there is often human agency involved.

As we study transportation, we think in terms of CLIOS, a broad characterization of the field.

Transportation System Concepts and Characterization

In our discussion of transportation system concepts, we will begin by using a simple taxonomy to look at how transportation systems can be characterized. From there, we will introduce the internal components of transportation systems. Then, we will discuss the external components of transportation systems: the components outside of the system itself. Using our taxonomy and building on our discussion of components, we will explore the mathematical abstraction of networks as a mechanism for describing transportation systems.

Next, we will analyze an elevator—a simple but rich example of a transportation system. The *elevator* system, although vertical in orientation, is a transportation system and a useful teaching tool. We will use this example to develop 30 key points that can guide us in the investment, operation, and use of transportation systems. We will close by discussing models and frameworks, and how we use these in transportation systems analysis and design. The outline is shown in Figure 1.1.

To begin with, we must create a taxonomy—a characterization of transportation systems. There are many ways one can characterize

FIGURE 1.1
*Transportation systems
concepts.*

Transportation system characterization
Components of transportation systems
   Internal
   External
Networks
A "simple" example
   Elevator system
Transportation "key points"
Models and frameworks

transportation systems. What we present (as shown in Figure 1.2) is not unique, but has proven useful.

In thinking about transportation, it is helpful to make a distinction between systems that focus on the movement of travelers and systems that focus on freight. While there are similarities in the ways these systems may operate, there are some fundamental differences in the characteristics of traveler and freight systems. On another dimension, we talk about the geographical areas that are serviced by transportation systems. We consider urban transportation and intercity transportation. Also, and quite importantly, particularly in recent years, we will consider international transportation systems.

Transportation systems differ in scale. People drive 10 miles to work but fly 400 miles to Washington and fly 10,000 miles to Japan. We send a letter across town, a few miles away. We move freight on an intercity level, several thousand miles, perhaps across the country. Many

FIGURE 1.2
*Transportation systems
characterization.*

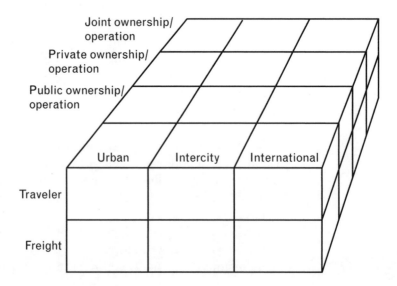

thousands of miles are involved in international moves that often require ocean transportation. Transportation has not quite reached an outer-space scale, but someday we will see that as well.

Another part of system characterization is how transportation systems are owned and operated. We distinguish between transportation systems that are owned and operated by private, profit-making organizations and systems that are provided by public sector organizations, such as federal agencies or regional and local organizations.

Often, we have a mixed, public/private type of enterprise. An obvious example is the United States highway system. The highway is publicly owned. However, a privately owned set of vehicles, owned and operated by trucking companies and people like us, use this public facility.

Transportation systems can also be exclusively private or exclusively public. The organizational context is critical in the operation of transportation systems.

Let us emphasize here at the outset, that there are many elements external to the transportation system. Customers, the most important of these elements, include travelers, shippers, and receivers of goods. We will focus considerable attention on customers and their needs later in this discussion.

In the next chapter, we begin our discussion of transportation system components.

REFERENCE

1. Sussman, J. M., "The New Transportation Faculty: The Evolution to Engineering Systems," *Transportation Quarterly*, Eno Transportation Foundation, Washington, DC, Summer 1999.

CHAPTER 2

Transportation System Components: An Internal Perspective

Physical Components of Transportation Systems

Having established an overall taxonomy, we can now talk about components of transportation systems. Our presentation here is not unique, as others might present transportation components in a different, but nonetheless useful way.[1] We begin with an internal perspective, focusing on physical components.

Infrastructure

We start with the notion of infrastructure, which is typically the fixed part of a transportation system. Infrastructure can be broken down into three categories: guideways, terminals, and stations.

There are a number of examples of guideways that we can identify, two of which are highways and railroads. Highways are general-purpose guideways—any rubber-tired vehicle can ride on it. Railroads, on the other hand, are special-purpose guideways in the sense that only railroad cars, with the proper spacing between wheels (gauge), can use the railroad infrastructure. We can also talk about guideways we cannot physically see in the way we can see a highway or a railroad, such as air corridors and underground pipelines.

---

1 Edward K. Morlok's *Introduction to Transportation Engineering and Planning*, McGraw-Hill, 1979, is an excellent general reference. Chapter 3 deals with the relationship between the transportation system as a whole and the requisite elements of the system.

Let's continue with the next category of infrastructure—terminals. Terminals are distinct from the guideway portion of the infrastructure. There are a variety of examples: terminals for buses like the one in downtown Boston, railroad freight yards, airports, and street corner bus stops in an urban bus system. Often serving a dispatching function, terminals regulate the departure of vehicles. Also, freight and travelers can enter and leave the system at terminals. Additionally, terminals often have a storage function. Terminals may be used to store empty rail cars or school buses during the summer or simply overnight. Finally, terminals serve as intermodal interchange points—places where people can change from the air mode in an airport to a taxi mode to go downtown, for example.

Stations are another example of infrastructure. When riding on a subway system, there are "stations" where you can enter or exit.

Vehicles

The next component, vehicles, is present in many, but not all, transportation systems. For example, there are no vehicles in a pipeline system. Unlike the guideway, the stations, and the terminals, which are fixed portions of the system, vehicles are components that move.

There are numerous vehicle types, including automobiles, trucks, railroad locomotives, and airplanes. In considering vehicles, their weight, durability, and crashworthiness are dependent upon the material out of which the vehicles are made. Often, there is an efficiency (lightweight vehicles require less energy to move) versus safety trade-off (lighter vehicles come out second best in a crash).

Consider automobiles, locomotives, and airplanes—each vehicle has its own method for propulsion. There are a number of vehicles, in various modes, that cannot move under their own power. Freight cars in the railroad industry, for example, have no power and have to be pulled by a locomotive. Containers or trailers in the trucking industry have to be pulled by a tractor unit in order to get from place to place. We talk about vehicles with propulsion and vehicles without propulsion as a way of distinguishing vehicle types.

Equipment

Another component is what we call "equipment." I put it in quotes because it is not always literally equipment. Equipment is machinery that operates to facilitate the transportation enterprise. For example, a loading crane at an intermodal terminal that takes a container off a flatbed truck and places it into the hold of a container ship is an example of equipment. Equipment for railroad track maintenance is another example.

Airport baggage handling equipment is subsidiary, but essential, to the traveler transportation function. Other examples of what we loosely call equipment include storage facilities for fuel, sand and road salt, as well as roadway snow removal equipment.

Power Systems

The next component is the power system. Power systems include the electric motors that propel vehicles. Internal combustion and diesel engines are examples of commonly used power systems. Power systems usually constrain the maximum speed and acceleration attainable. Some power systems use regenerative braking, in which energy expended in braking the vehicle is recaptured for later use.

Human power—using bicycles or walking, for example—is another form of power systems. Animal power is still a major power system in the developing world. Sailing ships, using the power of the wind, is a system with historical importance and is still in use for recreational purposes.

Fuel

Fuel is another transportation system component. Fuel may be gasoline, natural gas, diesel fuel, ethanol, methanol, coal, etc. The costs and efficiency of transportation fuels are of considerable interest. Also, the environmental impact of transportation systems is closely tied to the kind of fuel that is used.

Electricity is a fuel. The New York City subway system runs on electric power, as do intercity trains in Japan, France, and, to a modest extent, in the United States. Electricity is centrally generated—say, from coal—and transmitted via high-voltage lines to permit the propulsion of vehicles by onboard electric motors. Conversely, electric automobiles run on electric power that is stored in batteries that propel an electric motor.

Solar energy is a possible fuel for running automobiles with electric motors equipped with solar cells that convert solar energy to electricity.

The idea of having battery-powered automobiles, operating in conjunction with solar panels as a method to recharge the battery during operation, is a potentially important concept.

The internal combustion engine is the heart and soul of the worldwide automotive industry. Internal combustion engines (ICE) operating on gasoline are used in an overwhelming percentage of vehicles. As demonstrated by the oil shocks in 1973 and 1979, the geopolitical aspects of the dependence on foreign oil is very important in the United States. The scarcity of oil, as well as environmental concerns, motivates much of the research on other sources of vehicle power. Research and development is being done on hybrid vehicles, which have both an ICE and batteries that provide power to electric motors. The idea behind hybrid vehicles is that the batteries can take care of power surges during acceleration and thereby substantially improve gas mileage.

We do not often think of gravity as a source of power for transportation. Anyone who has ever ridden a bicycle or jogged, knows that it is easier to travel downhill than uphill. In fact, transportation companies take advantage of gravity as a component of the propulsion system of their transportation enterprise. The Norfolk Southern Railroad, for example, is a railroad that transports a lot of coal. The Norfolk Southern runs empty coal cars uphill to the mountains to get the coal, and they run full coal cars downhill to the port of Newport News, Virginia, for domestic use and international export. The Norfolk Southern takes advantage of the elevation of their coal source and uses gravity as power to run their full cars downhill to the port. Not only does this save energy, but it represents a good deal of money.

Control, Communications, and Location Systems

Other fundamental components of transportation systems are control, communications, and location systems. These components are various methods for controlling vehicles, infrastructure, and entire transportation networks. Often, this control component is in the guise of a human being, like the operator of an automobile, train, or airplane. This human control component may also be an air traffic controller, who is nowhere near the vehicles but is, in fact, controlling the system with instructions to pilots. A taxi dispatcher allocating passengers to particular taxicabs is a further example of human control in a transportation system.

In addition to the human elements of the control, communications, and location systems, there are elements of technology as well. For example, one could argue that traffic signals—red, green, and yellow lights—are an element for controlling the transportation system. Road signs (static or changeable) serve similar functions.

In some transportation systems, we have systems of control that are based on fully automated vehicles (vehicles that run without a driver). Many airports use fully automated vehicles to move passengers among terminals.

In order to avoid collisions, railroads have block control systems that prevent the entry of a train into a block while another train is in that block. Clearly, there will be a trade-off between block length, capacity, and safety.

A relatively new technology that is increasingly important in the transportation industry is satellite communication, and the Global Positioning System (GPS), as a location system for transportation enterprises. With GPS, one has the ability to locate individual vehicles and potentially aid drivers in their navigation task. For example, drivers would have a GPS sensor in their automobiles that would allow a central system to track the vehicles and make benign judgments (we hope) about how to route the drivers from origin to destination most efficiently.

Summary of Basic Physical Components

Let's summarize the basic components of the transportation system discussed so far (see Figure 2.1).

FIGURE 2.1
Transportation physical system components.

Infrastructure
  Guideway
  Terminals
  Stations
Vehicles
Equipment
Power systems
Fuel
Control, communications and location systems

Operators

Labor

Labor is the first of several different operators we need to be concerned about. Having already touched upon labor in the discussion of systems control, labor can be people that drive trains, pilots that fly airplanes, taxicab dispatchers, air traffic controllers, and automobile drivers.

In the category of labor, we include people who are concerned with system security, such as police officers that enforce drunk driving laws and speed limits, as well as control traffic. Other workers include maintenance and construction crews for all modes—maintenance and construction of infrastructure, vehicles, control systems, etc. Further examples are fare collectors on the Massachusetts Bay Transportation Authority (MBTA) and toll takers on the Massachusetts Turnpike. These people are not providing transportation directly, but are creating an environment in which transportation is safer, more secure, and more efficient.

We are seeing a trend in transportation where labor, to a certain extent, is being squeezed out by technology. For example, electronic toll collection is already deployed in many places around the world. With electronic fare collection, the concept is that a transponder in your vehicle is read by a sensor at the roadside, identifies you, and debits your account by the appropriate toll amount. Congestion is lessened because stopping to pay the toll is eliminated. However, from a labor point of view, the system implies fewer jobs because no one is needed to collect your money.

We are also seeing maintenance automation in railroad operations. In the last several decades, the railroad industry has gone through a substantial downsizing by replacing labor with technology. With electronic toll collection, fare cards for transit and lower manning requirements for ocean-going vessels, we can anticipate a continued downsizing of the labor forces in many transportation enterprises.

Organized Labor

Many transportation firms and agencies must deal with organized labor, often in the form of labor unions. In unions, workers band together to address issues of job protection, wages, and job safety. In the United States, labor unions grew out of extraordinary excesses of management power in the late 19th and early 20th centuries. The labor movement, of course, was not limited to transportation organizations. Coal miners and other nontransportation industries were also in the forefront of the labor movement. Since the labor movement, organized labor has been a major force in the history of transportation throughout the twentieth century in the United States.

Within the transportation industry, there has been a certain amount of labor strife. President Reagan basically put the air traffic controllers' union out of business in the early 1980s. The trucking industry deals with the Teamsters, one of the most powerful unions in the United States. The New York subway system has been subject to labor strikes from time to time. For many years, the labor contracts on the public transportation system in New York City expired on January 1st. People went to New Year's Eve gatherings without any real sense of whether public transportation was going to be able to get them home after midnight if the unions called the strike.

Many in management argue that labor unions have been a major source of inefficiency in transportation organizations and have led to situations in which particular firms cannot effectively compete for business. This lack of competition in the industry, as some argue, has led to negative impacts on the public-at-large. The term "feather-bedding" is used to describe labor unions using their muscle to assure unneeded jobs. The classic example of featherbedding is the inclusion of firemen on diesel locomotives. Although the firemen served an important function when locomotives were powered by coal-burning steam engines, they serve little purpose on today's diesel and electric trains, and these jobs have finally been eliminated.

Management Function

We now consider transportation management. Managers tend to be what Peter Drucker, the management giant, calls "knowledge workers." Managers are people that use knowledge and information in an informed way to advance the enterprise. Any number of functions fall under management.

Marketing

Marketing is a management function. There are people in virtually every transportation organization that are thinking about what their customers need, how their customers value the service they provide, and what the company should be doing to gain a larger share of the transportation market. Marketing is a fundamental function of any business and certainly a fundamental function in transportation.

Marketing is the area in which the dichotomy between public and private sector transportation organizations may be the strongest. This is a gross oversimplification, but in public sector organizations, the tendency over the years has been to fundamentally ignore the marketing function. "If people want to use the service, they will use it. If people do not want to use it, they will not use it. Our organization will survive."

On the other hand, a private company like American Airlines will spend a lot of money thinking through what kinds of services people want and estimating how much people are willing to pay for various services. Private companies also develop strategic approaches to building market share. In addition to other airlines, American Airlines might be competing with other modes, say, high-speed rail, for customers.

Competition Between Transportation and Communication

Airlines might also think about competing with communications systems, which might provide travelers with a new way of performing business that has traditionally required a face-to-face meeting.

A 1994 article in *The Boston Sunday Globe* discussed the telephone. The number of long-distance phone calls serviced by AT&T on a typical business day over the last decade went from about 37 million calls a day in 1984 to about a factor of five greater. In a mature industry that has been around since the 19th century, a growth of 400% in 10 years is noteworthy, and the growth continues at a rapid pace.

What is going on? People now have car phones and fax machines. People are using phones in a much more ubiquitous way. The notion of communication as a substitute for transportation is strongly illustrated by this increase in long-distance calling. How many trips were not made because of that factor of five increase in telephone calls in this country over the last 10 years? On the other hand, how much economic activity was generated by that level of communication which might have led to more travel? One could argue that communication reduces *or* increases the need for transportation. Currently, we do not have an answer for this question. The National Information Infrastructure (NII) is at the core of this new business opportunity. Telephone data illustrates that communication is a competitor, and marketing people within transportation organizations should think of it as such.

Strategic Planning

Strategic planning is another area that people in transportation management deal with. People in strategic planning worry about such issues as capital planning. Strategic planners ask questions like, "Should we build more infrastructure?" In a public sector organization, strategic planners might ask, "Should we build more highways?" In the case of a railroad organization, the question might be, "Should we build a new rail yard or should we build new trackage to provide service between particular origins and destinations?" Strategic planners have to make sensible judgments about how to invest in the transportation enterprise. In the transportation business—as in most businesses—investment decisions are complex. The estimation of capital investments requires strategic planners to make projections, in times of economic uncertainty, about future traffic in an environment in which they often do not know what their competitors are going to do.

Operations

Operations is another area of management. Operations management is concerned about how the system is operating today, from a customer service and operating cost point-of-view. Adding infrastructure is a strategic judgment. However, there are tactical judgments, such as bus scheduling and plane cancellations due to bad weather, that the operations managers make. There are a variety of operating decisions that people make to keep transportation systems operating in an effective and safe way.

Now, of course, there should be some effective interaction between these tactical operating people, the strategic people and the marketing people, in order to ensure appropriate investments are being made and services are being tailored to customers' needs.

Operations/Marketing Tension

Tension is a recurring theme in transportation. In the transportation industry, tension often exists between operating personnel and marketing personnel.

Now, when I use the word tension, I am not using it in a pejorative sense. This is business tension, possibly creative tension, not personal tension.

Marketing people like to provide high-quality service. To a first approximation, they want to maximize revenues—again, a gross simplification. The marketing people like to provide universal, direct, frequent, and high-quality service to transportation customers. The marketing people are basically concerned with maximizing the revenues that flow to the company.

The operations people are cost-oriented—again, a gross simplification. They are typically worried about minimizing cost. Operations people want to run an efficient and cost-effective operation.

To restate, there is a tension that exists between revenue maximization and cost minimization. This tension is not necessarily consistent with profit maximization in a private for-profit organization, as one group of executives trying to maximize revenues and another trying to minimize cost does not necessarily lead to optimal profits. But that kind of tension—the different perspectives of people who are trying to market high-quality services versus those who are trying to run the most cost-efficient system—exists in the airline industry, the trucking industry, the railroad industry, and in public transportation—in virtually every transportation organization.

Maintenance Management

Other management functions include maintenance—both the maintenance of infrastructure and the maintenance of vehicles are important management jobs.

Information Management

A sign of the times, part of the management function is information services. Some transportation organizations have a CIO, a chief information officer, concerned with assuring that the organization is taking proper advantage of information technology.

Operations Research

We also have, in many transportation organizations, an analytic quantitative operations research group that is concerned with providing support for operations planning and other parts of the organization.

Administration

Another kind of a catchall within the overall management framework is administration. Administration covers a variety of functions that are required to run any kind of enterprise, be it public or private. For example, transportation organizations will always have a legal department. Organizations in the private sector will always have executives who are concerned with interacting with the regulatory community in the public sector. In transportation enterprises, there will be people who are concerned with labor relations, as well as people in administration who deal with finance, since transportation organizations often have to go to public markets for capital.

Technology is making management jobs disappear—this phenomenon is not limited to labor. There has been a large downsizing of middle management in many transportation organizations.

All of the above is part of management. Some of the positions, roles, and functions have a special twist because of the world of transportation in which the organizations are operating.

Operating Plans

Now, in addition to the people, labor, and management that run the system, we also need operating plans. Operating plans include a variety of elements. Again, completeness will elude us, but let's explore some examples of what goes into an operating plan.

Schedule

An important element of an operating plan is a schedule. When do the trains leave? When do the planes leave? When do the buses leave? There are a lot of different attitudes in different industries about schedules. In some cases, schedule adherence is a very important characteristic of service. In other cases, schedules are decidedly looser, and are not adhered to very closely. So, there will be some variations, from industry to industry, in schedule construction and in the actual performance that is delivered with respect to schedule. As a generalization, schedule adherence in passenger operations is more important than in freight operations.

Crew Assignments

Another element of operating plans is crew assignments. How do you take crews—like airline crews, train crews, or bus drivers—and assign them to elements of work? I would suspect, if you asked the typical business operator, "Is that hard or is that easy?" the answer would be, "Well, it is easy in my business. I go into my factory, and I tell people to be there at eight, and I tell them to go home at four, and everything seems to work. Sometimes somebody is sick, so I have to get a temp or someone works an extra shift. It's not a big problem."

Their factory, of course, does not move around like vehicles in a transportation system. Crew assignments in the transportation business turn out to be hard to construct. Because labor is expensive, it is important that crew assignments be done well. A good deal of work goes into air crew schedules, considering the allowable number of continuous hours that an airline crew can work, monthly days off for airline crews, cost minimization, matching pilots with the planes they are certified to fly, and so on. Crew assignments turn out to be a difficult problem, mathematically and operationally.

There is comparable work on assigning crews in the public transportation industry. Again, it turns out to be difficult. In the public transportation industry, crew assignments are hard because people do not come to work with any high degree of reliability. The degree of absenteeism in the public transportation industry perhaps rivals that of any industry in the United States.

A further crew scheduling problem in the public transportation industry is the bimodal demand for transit service. A lot of people want to use transit from 7:00 to 9:00 a.m. and again from 4:00 to 6:00 p.m. to go to and from work, but from 10:00 a.m. to 3:00 p.m., ridership declines significantly. How do we assign drivers to a system that has this kind of bi-modal characteristic? One way is to ask them to work two shifts a day, the so-called "split-shift." For reasons that I suspect anybody who has worked for a living can understand, people do not usually like split shifts. Perhaps the split shift explains some of the absenteeism. How does one actually run a system that has this kind of demand profile in a cost-effective way?

Flow Distribution

In transportation systems, we often observe inherent imbalances in flow. For example, look at the morning peak and the evening peak. In the morning, everybody wants to come from the suburbs into Boston, and in the evening everybody wants to go the other way. There is a major flow imbalance. The flow averages out over the day, but there are substantial short-run imbalances.

We have this imbalance in freight systems, too. There are flows from the agricultural areas to the cities that are not reflected in flows of comparable magnitude from the cities back to the agricultural areas.

Those imbalances often result in the need to redistribute empty vehicles back to where they are needed. How one redistributes vehicles around the transportation system, recognizing that the traffic is inherently imbalanced, is an element of the operating plan.

Connection Patterns: Hub-and-Spoke

Another important function of the operating plan is the definition of connection patterns. One of the most common connection patterns is the "hub-and-spoke" network operation in the airline industry. In this kind of service, hubs are used to consolidate traffic—people are flown from several cities to a hub area where they are consolidated. The passengers from origins 1, 2, and 3, are consolidated into a single, perhaps larger aircraft, and are flown to a destination. From time-to-time, it is even possible that passengers end up "double hubbing," going through two hubs to get from origin to destination (see Figure 2.2).

FIGURE 2.2 *Hub-and-spoke operations.*

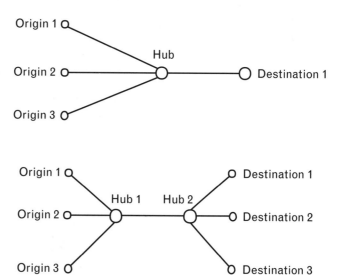

Although it looks easy on paper, it turns out to be rather compli-cated to run a hub-and-spoke operation. Operators have to consider connection patterns, the time required to transfer from one airplane to another, and the sensitivity of the system to bad weather. Bad weather at one of the hubs could negatively impact the entire system for long periods of time.

Connection patterns on the transportation network are an impor-tant element of the operating plan. The scheduling and routing of vehicles and people over the system is a fundamental issue.

Cost/Level-of-Service Trade-off

Transportation organizations are concerned about the level-of-service they provide. As Figure 2.3 shows, level-of-service is dependent upon connection patterns.

For example, do we provide direct, high-quality service from A to C as shown in the lower figure, or do we consolidate passengers at Node B with other passengers from Node D, into a single flight from B to C as shown in the upper figure?

Here we have some fundamental cost/level-of-service trade-offs. Which way do you think the VP of Marketing argues? What does he say when he talks to the CEO? "Hub them through B, no problem." Is that what he says?

FIGURE 2.3   *Two connection patterns.*

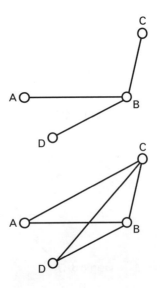

No.

Correct. He probably says, "Let's do direct service from A to C. The passengers will use our airline if we just provide direct service."

How does the VP of Operations respond? "Well, Charlie, it is kind of expensive to do that. There are very few people who want to take that trip, and it's just not cost-effective to provide direct service. Why don't we just take a small plane and hub them down to B, and consolidate them with the traffic from D? Then we can fly a decent load from B to C."

And the Marketing VP says, "Gee, Jane, if we do that, people won't fly our airline, because our competitor is flying direct service from A to C. The quality of service will be poorer on our airline, because we are going to ask A-C people to stop and transfer at B."

This kind of discussion happens all the time. We are dealing with the trade-off between cost and the quality of service to the customer. Is it better to provide a less expensive service with a lower price and a lower service quality, or a premium service with a premium cost, and most likely, a premium price? In operating a transportation system, the trade-off between cost and service is fundamental. The questions involved in this trade-off are often much more sophisticated than presented here; however, these provide a reasonable introduction.

Contingency Planning

One last aspect of operating plans is the notion of contingency planning. In contingency planning, the question is, "What do we do when things go wrong? How do we decide how to alter our operating plan to reflect changes in weather, demand for service and accidents—such as a derailment?" Because there are any number of things that can happen, developing an operating plan that is robust in the face of these uncertainties is important. Building a good contingency plan often requires an allocation of additional resources such as extra locomotives to buffer against high-demand possibilities. But extra resources are expensive. Providing flexibility is expensive and these expenses are likely reflected in the price of service. This is simply another example of the cost/level-of-service trade-off we mentioned earlier.

This completes our discussion of the internal components of transportation systems. The next chapter introduces the external components.

CHAPTER 3

Transportation System Components: An External Perspective

External Components of the Transportation System

Thus far, we have talked about the internal aspects of the system: people, equipment, operating plans, and the like. In addition to these internal components, there are components external to the transportation system—organizations and other entities—that have to be considered when thinking about the overall transportation enterprise. Figure 3.1 illustrates these external components.

FIGURE 3.1 *External components of the transportation system.*

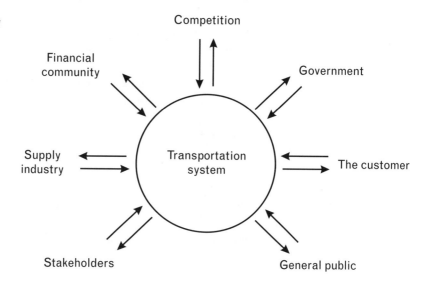

Government

The first entity is government. If you are a private company running a transportation service—a railroad, trucking company, and so forth—you and the government are partners to a certain extent (at the very least, you pay them taxes). Certainly, the government is concerned with the safety of your operation.

In some cases, the government will provide you with some of the infrastructure you need to run your system (usually for a fee). A trucking company operates on the government's highways. If you are an airline, you operate using publicly owned airports and a publicly owned air traffic control system. Of course, as noted earlier, the public sector itself may provide the service, as with most urban subway systems, for example.

The government is also concerned with individual transportation companies gaining monopoly power. Some of these concerns trace back to the late 19th century, when United States railroads held monopoly power. Now, a hundred years later, although the monopoly power of railroads has long since disappeared, some regulatory structure still exists. This is sometimes appropriate, as railroads still have captive shippers that have no viable transportation alternative.

Competition

Competition is a second element that is external to your transportation system. If you are an airline, you certainly worry about the other airlines that are servicing the same cities you are servicing. If you are a trucking company, you certainly worry about other trucking companies. At the same time, you might worry about other modes, such as an airline worrying about railroad companies. For example, if you are an airline serving the Northeast Corridor of the United States, you wonder about what will happen if high-speed rail service is actually deployed along the Northeast Corridor.

As noted earlier, methods of communication can be competitors to the transportation industry. Instead of flying to Chicago to conduct business, fax machines, Federal Express, and electronic mail (e-mail) can deliver information with high reliability at a fraction of the cost of flying.

Competition for the entertainment dollar is another example. A substantial fraction of travel in the United States is made for purposes of vacations and entertainment. If people decide they want to fix up their house rather than go on vacation, there is competition for that discretionary income that may reduce transportation demand. On the other hand, shipment of home improvement materials may go up.

Financial Community

A third external entity is the financial community. The transportation industry needs dollars for short-term tactical needs and long-term strategic needs. In the United States, many private transportation companies that are listed on the various stock exchanges worry about what happens to earnings in the next financial quarter and the implications for that on stock price. Businesses in the United States are often characterized as overly tactical and overly concerned with near-term financial results. Perhaps stereotypically, Japanese businesses are often characterized as focusing on long-term results and being less concerned with near-term profitability. If the characterization of the United States industry is correct, then decisions may be made to optimize short-term results that may not be effective in the long-term. Management in the United States often bemoans the lack of patient capital available to industry.

Supply Industry

A fourth external entity is the supply industry. There are a number of industries that supply transportation organizations. General Motors, Ford, DaimlerChrysler, Honda, and others can be characterized as part of the supply industry. They supply people like us with automobiles, which we, in turn, use for transportation purposes. Indeed, the relationship between financial health in much of the developed world and the health of the automobile industry is well documented and well researched. At MIT, "The International Motor Vehicle Program" looks at the relationships among economic growth, the health of the automobile industry, and how the industry can be more efficient [1].

Selling train locomotives to railroad companies, General Electric and General Motors are part of the supply industry. Energy companies that provide fuel, like Shell and Exxon, are also part of the supply industry. Likewise, Goodyear providing tires is a supplier. Boeing and Airbus are suppliers of airplanes to the airlines. Organizations that provide control technologies, computers, and equipment that allow transportation organizations to run complex global positioning systems or air traffic control systems are all suppliers.

I would even go so far as to characterize the research community as part of the supply industry. Organizations like MIT, Northwestern, and consulting organizations like Charles River Associates or Arthur D. Little could reasonably be characterized as suppliers of particular kinds of services to the transportation industry. The insurance industry is a major supplier to transportation organizations.

Stakeholders

In addition, we have a number of stakeholders in the transportation field. Stakeholders are organizations and individuals that may not be users of transportation or explicit suppliers of services or goods to transportation organizations, but are vitally concerned with transportation enterprises and their operating and investment practices.

An example of a stakeholder would be the environmental community. Lobbying to minimize the environmental impacts of transportation, the Conservation Law Foundation is a part of the environmental community.

Abutters to transportation infrastructure are also stakeholders in transportation operations. Airport noise is a critical issue around virtually every major airport in this country. People who live near airports are constantly lobbying for flight procedures—so-called "noise abatement" measures. To minimize noise, airplanes maneuver to avoid settled areas, particularly late at night. Nighttime curfews are another example. Of course, there is some tension between airlines and airports and these abutters. People living near airports are concerned about their quality of life impacted by noise. Airline companies concerned with safety would argue that some of the maneuvers their pilots are asked to make in the name of noise abatement reduce safety.

General Public

One could argue that the general public is, in a very real sense, a stakeholder in the transportation enterprise. There are relationships between transportation and economic development, quality of life, and national defense—all issues of great interest to the general public. So, to a greater or lesser extent, everybody is a stakeholder in transportation.

The Customer

The customer, in many ways, is the most important element external to the transportation system that we will discuss. We introduce this concept in the next chapter.

REFERENCE

1.   Womack, J. P., D. T. Jones, and D. Roos, *The Machine That Changed the World*, New York: Rawson Associates, 1990.

CHAPTER 4

The Customer and Level-of-Service

The Customer

The customer of the transportation system, be it the air passenger or the coal executive who is deciding which railroad to use to service his or her transportation needs, is fundamental to the transportation enterprise. Thinking about what is important to the customer, from the perspective of what services transportation organizations provide, is fundamental.

The customer is the person or organization that buys transportation services. Customers and their needs are fundamental to the existence of transportation organizations.

"Customer" is the buzzword of the 1990s management literature. For example, in Hamel and Prahalad's *Competing for the Future*, defining the market opportunity is fundamental to the transportation enterprise [1]. Drucker talks about all results being external to the organization, and the purpose of the business being the creation of customers [2].

If you are a transportation provider, what does your customer really want from you? A good way to think about your customer's needs is to think about what the customer's customer wants in the way of service.

Freight Transportation Customers

Freight customers are shippers and receivers of goods. The characteristics of these goods may vary quite substantially. For example, General Motors or Toyota ships automobile parts from all over the country to be assembled into finished automobiles. Once assembled, these finished automobiles are distributed to automobile dealerships, where they are available for purchase by the consumer— the freight customer's customer. In this particular instance, there is a process where the freight being transported—auto parts like fenders, windshields, and brakes—are assembled into a finished automobile. As with other industries, freight transportation is a central element of the automobile industry.

A recent thrust in freight transportation, particularly for manufactured goods, is supply chain management. Here, the transportation provider—the railroad, for example—becomes an integral part of the customer's logistics system, providing reliable delivery and pick-up of goods as they progress through the supply chain.

We also have very different kinds of freight—different than high-value "finished" products like automobiles and television sets. For example, we have coal companies that ship raw coal from a mine to wherever it is going—to a power plant, to a ship to be carried across the Atlantic or the Pacific, or to an end-user of the coal. Coal is a product substantially different from a finished automobile. It implies a different kind of service than you would want to provide for finished automobiles. Shipping finished automobiles around the country requires much more stringent security than one requires in shipping coal. Whereas damage to coal is a virtually unknown concept, damage to finished automobiles is a very serious issue in the automobile industry. Automobile companies are concerned about getting new automobiles to customers in perfect shape, and the care with which automobiles are shipped around the transportation network shows the companies' sensitivity.

Coal is a commodity: it sells on the basis of price. Other examples are grain and oil flowing through pipelines. Commodities tend to be relatively lower in value per unit weight, as opposed to high-value, finished manufactured goods like televisions, fax machines, and cars. Commodities tend to be homogenous—coal is coal to a first approximation. Coal companies are not usually concerned about which coal from their mine goes to which customer. However, automobile companies are very sensitive to making sure that the right set of auto parts goes to the right assembly plant. If General Motors sends Buick parts to their Pontiac plant, they have a real problem.

Depending upon the cargo, there are very different kinds of freight service. For example, for low-value commodities, transit time between origin and destination is a relatively minor issue to people who are shipping those kinds of goods. However, transportation cost is very important. Because the value of the commodity is low and it sells virtually on the basis of price alone, it had better cost very little to move it around because customers are shopping for the best price. For most commodities, transportation costs make up a significant fraction of the price.

For manufactured goods, the transit time may be very important, because there is a high-value item in the pipeline, and the carrying costs of that inventory are high. The reliability in getting manufactured goods to retail outlets is important because you do not want to stock out of time-sensitive goods (such as wedding dresses in June).

Some level-of-service variables for freight are:

- Price;

- Travel time;

- Service reliability;

- Availability of specialized equipment;

- Probability of loss and damage.

The same customer may have different freight needs at different times. For example, I (personally) am a freight customer: I use the conventional U. S. mail, I use overnight mail for situations in which I need high-speed and high-reliability mail service, and on certain occasions I use UPS or Federal Express. I also have the need to ship bulk—books or papers. What I require from each of these carriers in various situations is different.

Somebody very much like me may come to a different judgment about vendor selection because we may place different values on various attributes of the services. I may put a lot of emphasis on cost whereas one of my colleagues may be more concerned about reliability. As with any transportation service, there are differences between individuals looking at the same service attributes. So, different coal companies may make different judgments in the same way that a colleague and I may reach different conclusions because we value service differently.

Finally, as shown in Figure 4.1, we have various modes providing different levels-of-service. We are concerned with a variety of modes for movement of freight. Freight modes can be trucks or barges on

FIGURE 4.1 *Modes, services, and customers.*

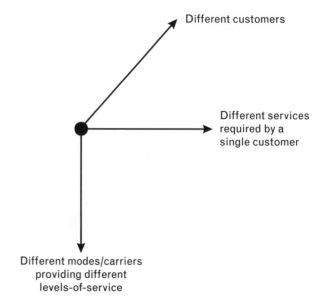

inland waterways like the Mississippi River and the St. Lawrence Seaway. Additionally, railroads and ocean liners provide transportation over very long distances at relatively low speeds and at relatively low cost. Air is also a freight mode; there is a lot of aggressive advertising for air freight systems, as various airlines try to establish themselves in this high quality (but high cost) mode. Finally, pipelines transport a large fraction of the ton-miles in this country—goods such as oil and coal slurry—but since we rarely see pipelines, people are not aware of them to the extent they are aware of the other modes.

Table 4.1 shows how ton-miles are split among modes in the United States.

Traveler Transportation Customers

Let us shift to the traveler side. Here we speak about urban traveler transportation, intercity traveler transportation, and international transportation. People travel for many reasons: school, jobs, social activities, medical needs, and so forth. Their choice of how to travel will often vary as a function of the purpose of their trip.

Travel modes are well known to you. In discussing traveler transportation modes, we can talk about privately owned vehicles such as

TABLE 4.1   U. S. Freight Transportation in 1993. (*Source: Transportation Statistics Annual Report 1994,* Bureau of Transportation Statistics, U.S. Department of Transportation, January 1994.)

MODE	% OF TON-MILES
Railroad	34.8
Truck	24.5
Water	21.9
Pipeline	18.5
Air	0.3

automobiles and publicly owned vehicles, such as buses and trains. In the United States, trains are a relatively minor mode for intercity travel, but an important mode for commuter operations in and around large urban areas. Trains are, of course, a very important mode in many other countries such as Japan and France, where high-speed trains are very popular. Additionally, we have air transportation for both long and short trips.

We have ship transportation, which is usually more for entertainment and tourism (e.g., cruises) than practical transportation. However, looking back to the early part of this century and up through World War II, the mode for crossing the Atlantic Ocean or the Pacific Ocean was passenger liner. While people still do take ocean liners, it is almost invariably for pleasure. Their business traveler market share is trivial.

Some important level-of-service variables for traveler transportation are:

- Price;
- Travel time;
- Service frequency;
- Comfort.

Choice and Level-of-Service for Freight and Travelers

Customers, be they people who are shipping freight or people who are making judgments about traveler modes, usually (but not always) have a choice. They have a choice among modes and they have a choice about carriers within a particular mode. Indeed, this choice is fundamental to our approach. The choice a customer makes will presumably be in their own self-interest. They will optimize some set of variables by studying the level-of-service for various options and choosing the one they find most acceptable.

Level-of-service is a complex concept. In transportation, level-of-service is multidimensional. Level-of-service is not simply measured by one variable like travel time or safety. Rather, it is some combination of a number of different variables that people—travelers, shippers, receivers—will somehow integrate and internalize in order to make a judgment about what mode and carrier to select. Individuals who seem to be similar may make different judgments about which modes and carriers to use. Customers who have different needs make very different decisions. You would not expect General Motors to use the same attributes of level-of-service to make a judgment about choice of modes and carriers as a power plant receiving coal.

The choice of mode and carrier may change for a customer over time—people change their minds or circumstances change. Currently, each of us is in a particular financial strata, and this financial strata influences our transportation decisions. So, people change, both in the long run as their circumstances change, and as the particular situation changes. For example, if I have been a little slow in getting my proposal out, and it is due in Washington tomorrow, rather than a week from tomorrow, I may, dynamically, choose express mail over some less expensive mode because I "absolutely, positively" have to have it there tomorrow.

Level-of-Service as Dynamic

The level-of-service provided by the carriers will change—and sometimes change dramatically. One should not think of level-of-service as static, as in fact, it is quite dynamic. Perhaps most interestingly, the level-of-service provided by the carrier will change as a function of the volume being carried by that carrier. So judgments made—not only by you, but by me and everybody else—will affect your level-of-service.

Putting it simply, as the volume, say, on Route 2, west of Boston, increases at Fresh Pond at 8:00 a.m., the level-of-service to me, as measured in travel time, changes. My choice to take Route 2 affects the level-of-service to a certain extent, but what I am really affected by is that a lot of other people have chosen Route 2 as well, and my level-of-service deteriorates. Sometimes, it may work the other way. If more people want to use the Number 1 bus on Massachusetts Ave., perhaps the MBTA will increase the frequency of buses (perhaps not) and my level-of-service will improve.

So, we are talking about a highly dynamic situation where the level-of-service is changing, people's needs are changing, and level-of-service is multidimensional.

It is important to recognize that different organizations may play different roles in different parts of the transportation process. For example, consider that I am UPS's customer and I call them at 2 o'clock in the afternoon and say "I need you here by 5 p.m. to take this overnight letter to Jacksonville, Florida." In this scenario, I am the customer and UPS is the provider.

At the same time UPS is a major customer of the railroad industry, and provides the railroad industry with enormous revenues, as my package and your package and many other people's packages are batched into a container that may go by railcar. In fact, UPS is one of the largest customers of the U. S. rail industry. So UPS is both a carrier and a customer—in this case, a customer of the railroad industry. When the railroad industry thinks about the service it is providing, it has to think not only what its customer wants (in this case UPS), but what the customer's customer wants (me, or other UPS customers).

Now, just as I have choices—Should I use UPS? Should I use Federal Express? Should I send a 36-page fax? What shall I do?—UPS has choices as well. Should it use rail? If rail, should it use CSX or Norfolk Southern? If not rail, what? Truck? If truck, Yellow Freight or Consolidated Freightways? How about using air as a mode? In the case of UPS, they have decided to be an air carrier as well. They made a strategic choice to "own," not "buy" air services. Clearly, there are many choices.

In transportation systems, the ultimate consideration is the customers—people or organizations that use transportation services. To some of us, this may seem rather obvious. However, in the real world of transportation, it is remarkable how often the customer is not the first priority, and may be low on the list when looking at how the systems are operated.

You have all, I am sure, run into individual employees of transportation companies who you would argue are not particularly customer-oriented. We have all encountered a nasty clerk at an airport, on Amtrak, or at the post office. But even as one goes up the management chain, the concern with operations and keeping the system running efficiently dominates thinking about what the customer needs. The old joke about, "We would be better able to keep our buses on schedule if we didn't have to stop for passengers," is too often a management attitude. So, I belabor the point only because customer service is often overlooked, even in a management era in which the customer is paramount.

Aren't schedules driven by customer demands?

Often not. Schedules are loosely a function of customer demands. That is, there are more trains going from Lincoln to Boston during the morning rush hour than at noon, but schedules are also affected by operating efficiency and cost control considerations. Customers are a factor, but sometimes they are a modest factor. The rate at which, for example, commuter train schedules change in response to customer demands is modest. I take the train (in moments of desperation) from Cambridge to Lincoln. In my briefcase I have the schedule I obtained five years ago. The schedule hasn't changed at all. It is hard to believe that the customer needs for service going to the western suburbs has not changed at all over that period.

Well, I think that schedule changes are dependent upon the mode. In the airline industry, schedules change all the time—the airlines are quite service sensitive.

Indeed. Distinctly different from public transportation, the airline industry is a private profit-making organization. That is, they plan to make profits, but don't always do so. Amtrak is pseudo-public and a commuter rail is almost always a public operation, so the sensitivity to customers may be different. By the way, the airline industry has taken, as an article of faith, that mode share in a particular corridor, say, between Boston and Chicago, is directly and tightly tied to service frequency. That is, if you are an airline providing 60% of the seat miles between here and Chicago, you're going to get about 60% of the traffic.

Airlines may be less service-driven than you think. Hub-and-spoke service is far from optimal from a service point of view.

Yes, an excellent point, and one we touched on earlier. The fundamental trade-off the airlines and railroads make in not providing direct service is a cost/level-of-service trade-off. Of course, airlines argue that they can't afford direct service between certain cities because there is insufficient demand. Perhaps if they provided direct service their fares would need to be higher, which is not good from a level-of-service viewpoint.

Well, I think it may be more productive to think in terms of what customers value. For example, Federal Express was able to figure out that people were willing to pay more to guarantee that their packages would be delivered at a certain time. So there was a sufficient market to justify the investment and development of that system. The problem frequently, with most of the public systems you mentioned, is that I am not given the choice as the customer willing to value something more.

Indeed. You are illustrating the complexity of level-of-service, and the fact that it is valued differently by different people at different times. Federal Express, incidentally, also determined that people put a high value on information about their shipment. People want to buy the reliability and the travel time, but they also want to buy information. Knowing when their parcel is going to get there and knowing if it has gone astray is valuable information.

Reducing Multidimensional Level-Of-Service to a Unidimensional Variable

Now, level-of-service is clearly multidimensional; however, for the purposes of analysis, we often need to reduce it to a unidimensional variable.

Level-Of-Service Variables

Consider air transportation between Boston and New York. What are the level-of-service variables that you would consider to be important? What measures would you, as a passenger, use in evaluating this service?

Travel time.

Travel time, of course. The time we spend in the air is very important.

What about the type of aircraft?

Fair enough. I do not really like to ride those 12-seaters. I would rather ride in a larger jet aircraft.

Just to make it simple, we could talk about what I call "comfort." Comfort includes factors like the kind of airplane and whether the flight attendants have a reputation for high-quality service. We could even choose to bundle "safety" into the comfort index.

Frequency of flights.

Okay, frequency of service. That is a reasonable level-of-service measure.

From frequency of flights, can't you simply talk about average waiting time?

We could do it that way. We already have an air travel time parameter. What other kinds of times are there?

Access time.

Good, it takes time to access the airport. It takes me, from the Western suburbs, maybe 45 minutes, depending on time of day, to get to Logan International Airport. But in addition to access, service frequency relates to waiting time. Suppose there is hourly plane service between Boston and New York. It probably isn't reasonable to assume that the average waiting time is a half-hour. I do not think people generally arrive uniformly over that hour. If I'm flying, I know what time the flight is supposed to leave. So, depending on my own preference, I try to arrive X minutes before the flight, where X varies from person to person.

Now, on the other hand, if I have a bus service that has a service headway of every five minutes, it is probably reasonable to expect that people arrive uniformly. People do not try to catch a particular bus. The average waiting time can be reasonably assumed to be half the interval between buses.

We introduced "waiting time" which could include time waiting in the airport for the plane to take off. It might include waiting for your luggage at your destination. It might include waiting in the queue at the cab stand at LaGuardia. Anything else?

The fare is important.

Yes, in evaluating level-of-service, we will consider the fare. We are more likely to take a service if it cost $50.00 rather than $125.00. This is clearly a level-of-service variable.

We have some good variables here—but what about units for each variable? Travel time in the air or on the ground and waiting time is measured in time—say, minutes. Fare, of course, would presumably be in units of dollars, yen, or francs.

Comfort—we will have to come up with something. "Hugs" could be a unit of comfort. Measuring comfort in hugs is an arbitrary way of combining ticketing efficiency, safety, and other factors like seat width and food service.

In Table 4.2, we have several variables and units.

The idea, of course, is these units are different. Hugs are different than dollars, which are different than minutes. But we need to measure the level-of-service in a unidimensional way. We need a way of collapsing these variables into a single variable.

Utility

So, we define a measure called "utils"—for "utility." And what I will do in the next step is convert all of these level-of-service variables, measured in different units, to utils.

TABLE 4.2   Level-of-service variables.

	VARIABLES	UNITS
Travel time	t_t	Minutes
Access time	t_a	Minutes
Waiting time	t_w	Minutes
Fare	F	$
Comfort	H	"Hugs"

Let's define a variable V, which is the "utility" of a traveler's choice in utils:

$$V = a_0 + a_1 t_t + a_2 t_a + a_3 t_w + a_4 F + a_5 H$$

We select the coefficient $a_0, \ldots, a_5$ so that the units of all of the terms are the same—utils. So, for example, a_0 is simply in utils; it is a constant that we would use to calibrate the utility equation. (We will not get into the discussion of how that calibration takes place here.)

a_1 would be in utils/minute, which would be a measure of how people value their travel time, which may vary from person to person. For example, people that make more money may tend to value their time more highly than people that make less money.

a_2 would also be in utils/minute, and might be different than a_1. People may value access time differently than they would travel time.

The units of a_3 are also in utils/minute, and may have a different value than a_1 and a_2, although all relate to time.

a_4 has units of util/dollar. And finally, a_5 has units of utils/hug—our catchall comfort variable. And again, a_5 may vary substantially from person to person.

For now, we are not dealing with how one would actually come up with a_0, a_1, a_2, a_3, a_4, and a_5. Let us talk about these terms just to make sure we fundamentally understand what we are doing. We assume that utils are good—that is, the more utils you have, the better you perceive the service to be. Consider the numerical sign for each of the "a" coefficients. What sign would you expect for a_1?

Negative.

Yes, the coefficient would be negative. More time is bad, unless you really like airplanes and you want to stay up there as long as you possibly can. Most people would just as soon get from New York to Boston more, rather than less, quickly. The same could be said for a_2. Therefore, a_2 would be negative, as would a_3. Most people do not like to wait around at the airport. Now, there are exceptions, but in general I think one could argue that coefficient would be negative.

Comfort—what would you expect? The comfort coefficient would be positive. The more hugs you receive, the happier you are, and so that coefficient would be positive.

How about fare? You would expect that to be negative. The more expensive the service, the less valuable it would be for you.

....
Mode Choice

Suppose we have three possible modes of travel from Boston to New York—air, train, auto—and you could measure each of the level-of-service variables for each mode. We could compute the utility of each mode,

$$V_{air}, V_{train}, V_{auto}$$

and you could assume that the mode with the highest utility is the one you would choose. However, much modern literature in utility theory uses a probabilistic approach. For example, the probability a traveler selects the air mode is as follows:

$$P(air) = \frac{V_{air}}{V_{air} + V_{train} + V_{auto}}$$

or perhaps:

$$P(air) = \frac{e^{V_{air}}}{e^{V_{air}} + e^{V_{train}} + e^{V_{auto}}}$$

This probabilistic approach is intended to reflect the fact that people have different utilities, and perhaps we have not captured all the level-of-service variables in our formulation.

If we know the overall size of the market, we can approximate the mode volumes by multiplying market size by the probability a particular mode is selected.

So, we can now reduce a multidimensional level-of-service variable to a simple variable—in this case, we call it "utils." In this simple case we used a linear model to compute utility. There are multiplicative models, logarithmic models, and others. You will see more complex functions in more advanced treatments of this material.

Travel Time Reliability

Let us talk about another important level-of-service variable. In transportation, we talk about the concept of reliability—here I do not mean reliability in the safety sense (concern with the vehicle encountering some difficulty and crashing), but rather, reliability in the variability sense. There is variability in travel time. How do you factor variability into your decision making?

Here's an example. I have a plane to catch at Logan Airport and I'm leaving from MIT. It is 5:00 in the afternoon, and I plan to take a cab. I rush out to the cab stand on Massachusetts Ave. and say, "Please take me to Logan." The cab driver says, "What time is your flight?" I say, "5:45." (It was actually 6:00 p.m.—but I wanted him to be really aggressive.) The cab driver says, "We're in big trouble." I say, "Well, do your best."

The cab driver then presents me with two choices. He says, "There are two ways to go. We can go the standard way through the tunnel. That is a chancy route at this time of day. Sometimes it's empty, sometimes it's packed. I do not really know what it is like today. Like I said, travel time varies a lot, but on the average it takes thirty minutes. You might get lucky. It might take only twenty. You might get really unlucky, and it might take an hour and a half. I really do not know what it is going to be today." Figure 4.2 shows how it looks as a probability density function.

Figure 4.2
Tunnel route.

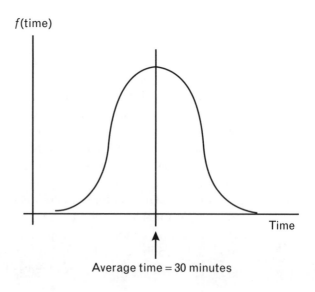

Then he says, "Here is your second option. I will take you through Charlestown, and we never go through the tunnel. And we go up-hill and down-dale, a rather circuitous route. It is a much longer distance," he says, "but I can predict with great certainty how long it will take, because there isn't any traffic there. It is just circuitous, because we are not going through the tunnel." Figure 4.3 shows how the second option looks as a probability density function.

Now, if we were to compare these two distributions, the average travel time going through the tunnel is lower than the average travel time avoiding the tunnel. So, if we were making the decision based on average travel time, we would presumably pick the tunnel route. But perhaps we would not. If we had a plane to catch we might say, "If I am virtually assured of getting to that plane by taking the higher average but lower variation route, that is what I'm going to do." So, I chose the more reliable—lower variation route—higher average travel time route, because it virtually assures me making my air connection, even though as it turned out, the cab ride probably cost about $4 or $5 more than it would have if I had chosen the tunnel route.

We will call variability in travel time "service reliability." In particular systems, service reliability in travel time can be as important as average travel time. Railroad and truck systems for moving freight differ both on a reliability dimension and on an average travel time dimension. Logistics and inventory theory help us develop an understanding of why, in economic terms, people might value a higher travel

FIGURE 4.3
Charlestown route.

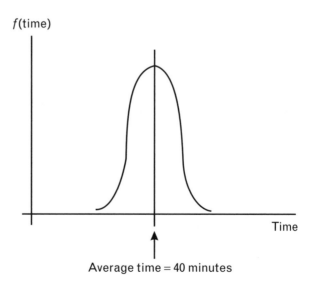

f(time)

Time

Average time = 40 minutes

time with better service reliability (i.e., lower variation) was discussed in Chapter 12.

That completes our treatment of customers and external components impacting the transportation system. We continue with an introduction to networks.

REFERENCES

1. Hamel, G. and Prahalad, C. K., *Competing for the Future*, Boston: Harvard Business School Press, 1994.

2. Drucker, P., *Management: Tasks, Responsibilities, Practices*, New York: Harper & Row, 1974.

CHAPTER 5

Networks

Networks

We now introduce transportation networks. Networks are a very convenient way of thinking about, characterizing and modeling transportation systems. In fact, the concept of the network is so basic that we have already used "node and link" diagrams several times to talk about transportation networks, without formally introducing the concept (see Figure 5.1).

Links

Links are typically guideways, highways, rail lines, air corridors, etc. We have links that can take flows, typically of vehicles, in one or both directions.

We describe transportation networks as interconnected. We have connections between the links through the other basic network element that are called *nodes*. Nodes often represent terminals or stations. (We will discuss nodes in more detail later.) In most real world transportation cases, the network is redundant. There are usually multiple ways to travel between nodes. There is often, but not always, an

FIGURE 5.1
Node and link network representation.

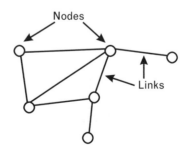

implicit link capacity. For example, one may define the capacity of a link as some number of vehicles per unit of time. We can think of capacity as a link volume beyond which the travel time is infinite (see Figure 5.2).

There is often a cost associated with traversing a link. The cost may be in terms of a travel time or there may be some explicit cost, such as a toll on a highway or an environmental cost. We may have a circumstance in which the cost, travel time for example, is a function of the volume being carried on that link. We can draw a relationship between travel time and volume on a link consistent with your intuition about highway performance (see Figure 5.3).

Now, Morlok talks about transportation networks as hierarchical [1]. He characterizes various levels of links. For example, in discussing urban highway transportation, he describes this hierarchy beginning with local streets—maybe a street in a residential subdivision. The local streets would then empty into somewhat larger collector streets. These collector streets would then empty into arterials. Finally, arterials would empty into a freeway or an expressway.

You can make similar characterizations for other modes. In rail, there is a high-capacity, well-maintained mainline track. Then there are more modestly maintained, lower capacity branch line tracks. At the bottom of the hierarchy, there are shipper sidings.

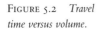

FIGURE 5.2   *Travel time versus volume.*

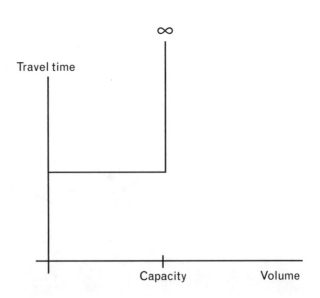

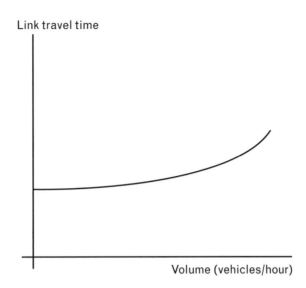

FIGURE 5.3
Link travel time.

Intermodal Networks

Networks are also useful for representing intermodal operations. For example, you could have a network that has roads going into a train station or a transit terminal. You would park your car and enter the transit system, such as at the terminus of the MBTA red line in North Cambridge. With some changes, the transit service could take you to Logan Airport. You then go onto an air link to some other airport, say Chicago's O'Hare, and reverse the process. In Chicago, you take transit, in this case with no changes, from Chicago's O'Hare to Chicago's downtown loop.

As shown in Figure 5.4, we see that we can represent a complex intermodal transportation system as a network. The vehicles are different on each link, as is the physical nature of the link itself.

Nodes

Now, moving on from our discussion of links, let us discuss the concept of nodes. Nodes often represent physical places:

- A terminal yard in a railroad operation;

- An airport;

- A parking lot.

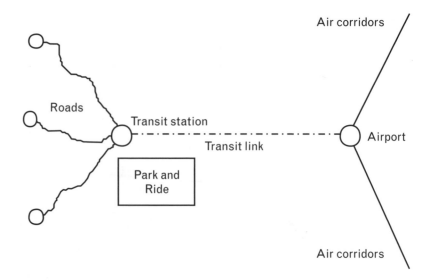

FIGURE 5.4
Intermodal network.

There is often a one-to-one correspondence with nodes and physical facilities (as there is with links). From time to time, we use nodes to represent a place in which the characteristics of a link change. For example, if a highway had two lanes and became a three-lane facility with higher capacity, you might model the capacity change by using an intermediate node. This intermediate node would demarcate a different kind of link at a three-lane section of the physical system than there is at the two-lane section (see Figure 5.5).

FIGURE 5.5   *Node to denote link change.*

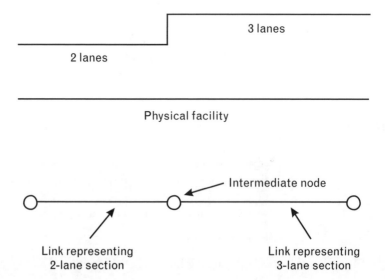

Think of networks as vocabulary. It represents a way of talking about transportation systems. Further, it neatly lends itself to a mathematical formulation of transportation systems.

Mathematical Operations on Networks

To motivate the notion that networks are important, let's discuss what kinds of mathematics you might want to perform on a network of the sort shown in Figure 5.6.

Predicting Link Flows

You may, for example, have an origin-destination (O-D) matrix for a particular network (see Figure 5.7). The O-D matrix would simply have the node numbers listed sequentially across the top and down the side. Each matrix entry would represent our prediction, somehow derived, of the flows (f_{ij}) or "demand" between nodes i and j. The main diagonal will be zero by definition.

A classic problem is: "How are the flows f_{ij} distributed on the network?" We are interested in the flows on individual links. Why would we care about what the flows on the links are going to be?

Because we need to design the size of the link.

The size of the links should be a function of volume. So, to come up with link flows we have to assign each O-D flow (or demand) to paths made up of connected sets of links.

Often in transportation planning, we choose the shortest path between origin and destination, reflecting an assumption that everyone wants to minimize travel time. This requires us to compute the shortest

FIGURE 5.6   *Typical network.*

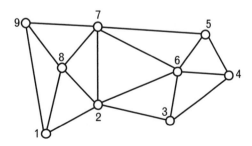

FIGURE 5.7
Origin-destination matrix.

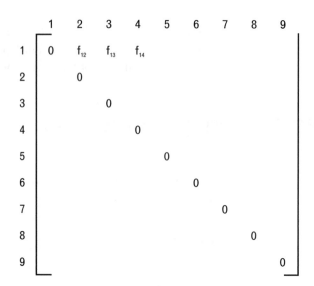

time path between all node pairs on the network. We start with some estimates of the travel time on each link, and compute the shortest path between every pair of points. So, for every node in the system you compute a shortest path to every other node in the system. A good deal of research has addressed this problem. It is a classic problem in transportation networks.

Now, let us suppose we have computed the shortest path between every pair of nodes on the network, and now we want to assign the origin-destination flows to the links. Now, one way of doing it is called "all-or-nothing" assignment. Referring to Figure 5.6, suppose we have 19 units that want to go from 1 to 3. We compute the fastest way to get from 1 to 3 (let's assume [1,2,6,3] is the shortest path) and we put those 19 units of flow on that shortest path.

We would assign the 19 units to all of the links between origin and destination, in this case link [1–2], link [2–6], and link [6–3]. Then we would repeat that again and again for all origin-destination pairs. We would conclude with all the O-D flows on the network having been assigned to links.

When we assign flows to a link, then do we change the link times?

Exactly right. First of all, we may assign much more to a particular link than a link can carry. The capacity of link [2–6] may be 16, using the definition of capacity shown in Figure 5.2. The link comes to a grinding halt if we assign more than 16 units to it; therefore we cannot possibly have 19 units on the link.

So, perhaps we would assign only 16 units to that path and then look for the next shortest path between nodes 1 and 3, until no more flow can be assigned. Then, we go to the next pair of nodes, until we complete the network. You should realize that the predicted link flows would be dependent on the order in which we did the origin-destination pairs.

But, more subtly, the performance of that link may deteriorate gradually as traffic travels over it. This is shown in Figure 5.3. Even if the capacity of this link was 50, not 16, the travel time on it when it was empty may be different than the travel time on it now that it is loaded with 19 units. So, at the very least, after we finish doing the node 1 to node 3 assignment, perhaps we should change the travel times on those links. Then we need to do the shortest paths all over again, because the shortest paths may now very well be different.

Incremental Assignment

An approach that can be used to deal with the kind of problem we just identified—that is, overloading the links, and the predicted link flows being dependent on the order of O-D assignments—is called *incremental assignment*. Incremental assignment is an attempt tonumerically solve the complicated problem of assigning O-D demand to the network.

With incremental assignment, we begin with an unloaded network and we have the travel times on all the unloaded links. First, compute all the shortest paths between all pairs of nodes. Then, what we can do—and this is one of many ways—is randomly generate the pair of nodes to assign. Using a computer, we use a random number generator to select a pair of nodes, say node 5 to node 2. Using the minimum path between 5 and 2 we assign a fraction of the traffic. Perhaps we choose to assign 20% of it to the links on the shortest path. Then, we recompute all the shortest paths again, because we have changed the state of the network. We then randomly generate another pair of nodes. Finally, we have most, or virtually all of our flows assigned, and we have a numerical solution to this particular network assignment problem. By randomly and incrementally assigning the flows, we get a better numerical solution than in the all-or-nothing case.

Again, a good deal of research has been done on how to solve these kinds of networks, by numerical methods and by direct optimization methods. The purpose here is simply to give you a sense of the kinds of issues to consider in using network formulations and highlight why networks are useful when thinking about transportation problems.

The Inverse Problem

There is another network problem called the *inverse problem*. In a real-world transportation network, we may not know the origin-destination demands—such data is often very hard to obtain. What we often do know, however, are the flows on the links (we can actually go out on Route 2 and see how many cars pass Fresh Pond between 8:00 a.m. and 8:30 a.m.). So, given the flows on the links of the network, the inverse problem involves computing what the origin-destination matrix looks like.

Logical Links

Finally, despite our use of examples where there is a relationship between physical elements of the transportation network and links and nodes (a highway is a link, an airport is a node), this is not always the case. In fact, we can use the concept of network modeling where links and nodes do not represent the physical system. We could construct a network that is not one-for-one with the transportation network itself. For example, a link could represent a freight car coming into a terminal on Train No. 1 and leaving the terminal on Train No. 7 some number of hours later. That connection could be represented by a "logical" link (see Figure 5.8.).

FIGURE 5.8   *Using a link as a "logical connection."*

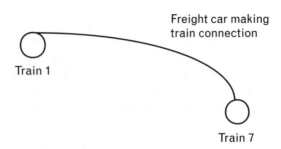

Freight car making train connection

Train 1

Train 7

This concludes our discussion of networks as a useful element of transportation vocabulary and as a valuable mechanism for analysis.

REFERENCE

1. Morlok, E. K., *Introduction to Transportation Engineering and Planning*, McGraw-Hill, 1978.

CHAPTER 6

Transportation Systems: Key Points 1–10

....

The Elevator Example

We are now ready to use an elevator system as an example of a simple transportation system, through which we will illustrate some key points in transportation systems.

I should mention that the invention of the elevator had a great deal to do with the development of cities in the latter part of the nineteenth century and the early part of the twentieth century. The development of elevators, paired with new technology to build high-rise steel skyscrapers, allowed a concentration of activity in urban areas that was not previously possible.

Elevators are simple compared to some of the more complex transportation systems, but they can be instructive and illustrative. With this simple example we can gain insight into overall system behavior that we can apply to more complex systems.

Elevator System Configuration

Let's assume that we have a 60-floor office building with three elevators. Each is configured to go from the bottom to the top—from the first to the sixtieth floor (see Figure 6.1). For the sake of simplicity, accept the three elevators as given. The actual number of elevators would be a design variable for this building.

Let us think about how we might operate the system. An obvious way to operate the system, although not particularly clever, is simply to have all the elevators provide service to all 60 floors. We would have three elevators, and every one goes from floor 1 to floor 60. From the perspective of a customer or passenger, you would walk

FIGURE 6.1
Elevator system.

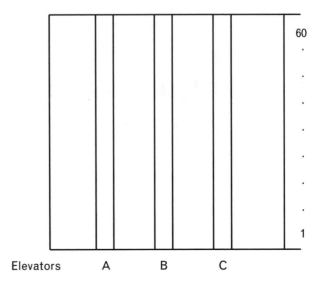

FIGURE 6.1
Elevator system.

into the lobby and regardless of what floor you want to go to, you could get into any elevator because all elevators go to all floors.

Now, that is not a terribly good way to operate the system, because the elevator is a sequential system. This is simply a sophisticated way of saying that if we are sharing an elevator and I want to get off on the twenty-fifth floor and you want to get off on the fifteenth floor, I have to wait while the elevator stops for you. So, the sequential nature of the system has some effect on level-of-service. The fact that I need to wait if I am going to a high floor for a lot of other passengers to be served suggests different ways of operating the system. We will discuss other ways of operating the system later.

The Elevator Cycle

Look at this from the point of view of the "elevator cycle." (We talk about vehicle cycles in many transportation situations.) Suppose the elevator sits there on the first floor. At some point it is "dispatched." Presume that the elevator is centrally controlled by a system that tells the elevator it is time to leave. Possibly, the control system uses some sort of dynamic algorithm. It might leave every minute, it might leave when the weight reaches some certain amount, it might do many things depending upon how sophisticated the system is.

The elevator would then go up to the highest floor for which it had a customer, stopping to let off customers along the way. The elevator would dwell there or it might continue up to the sixtieth floor—even though it has nobody in it. Eventually, depending upon what is going on with the other elevators, the control system will tell it to come back down. It will pick up people on the way down and will eventually get to floor 1 and start the elevator cycle again.

Now, before we go any further, I should note that the elevator system at the Volpe National Transportation Systems Center in Cambridge, Massachusetts, is my inspiration for this example. If you've been there, you know it is an interesting system—especially if you like to watch flashing lights on displays. There are four elevators with a display board that indicates where each elevator is, on what floors people have demanded service, and in what direction they want to go. So, we can look at the board and it will indicate that on the seventh floor, there is somebody that wants to go up. We do not know how many people are waiting on the seventh floor or where they want to go.

For people who like systems of this sort, you might want to go over to the Volpe Center. When you go over there, you will almost unquestionably have a lot of time to watch the lighted display, because the elevators are slow. I do not know how many people worked there when the building was first opened in the late 1960s; it is presumably less than work there now. I go over to the Volpe Center a fair amount, and I've learned to leave some slack time before my meeting because I usually have to wait a few minutes for the elevator. (I trust my friends at the Volpe Center will take this light-hearted example in the spirit in which it is intended. Indeed, some of the elevators at MIT [e.g., Stratton Student Center] are much worse!)

In most cases, we usually don't worry too much about wait time. Because elevator service is so reliable and frequent and the elevator time is small compared to the overall trip time, we don't worry about waiting for the elevator (except at the Volpe Center). Because the time I wait for, and ride in the elevator is a nontrivial fraction of the time it takes me to walk to the Volpe Center from nearby MIT, I alter my patterns. This is an example of how the behavior of the customer, in this case me, can be affected by the transportation system.

This illustrates the first of a number of *key points* we will develop in this simple example—we argue these key points are relevant and fundamental to understanding transportation systems.

Key Point 1:   Behavior
1. People and organizations alter behavior based on transportation service expectations.

How else might I alter my behavior? Suppose I am in the hall and I see the elevator about to leave. If I know the time between elevators is short, I might simply let the elevator go. If I know I will have to wait a long time for an elevator, I may yell to the people in the elevator to "hold the door." And, if they do, this further exacerbates service problems for customers waiting on other floors since the dwell-time of the elevator will be longer.

Now, does the level-of-service of the elevator system at Volpe matter? To an extent, it depends on who is doing the traveling. If I am going over there to try and get some money to fund transportation students, I am unlikely to complain or stop visiting Volpe. If I decide not to make the trip and some students are not funded as a result, do you think I am going tell the president of MIT I did not go over to the Volpe Center because the elevator service is so poor?

The point is that I am not walking over to the Volpe Center and riding the elevator for fun. It is part of a broader purpose—in this case, obtaining resources for students. This is essentially always the case. The transportation service is part of a broader system—getting to work, supplying goods, etc. This is our next key point.

Key Point 2:   Transportation as part of a broader system
2. Transportation service is part of a broader system—economic, social, and political in nature.

Returning to the Volpe Center, we ask why the service is poor on those elevators. A fundamental issue here is that I have no real option unless I want to climb 12 floors of stairs to keep my appointment. The elevator has no competition (and the Volpe Center has no local competition as a source of funds). This suggests our next key point:

Key Point 3:   Competition
3. Competition (or its absence) for customers by operators is a critical determinant of the availability of quality transportation service.

All elevators going to all floors sounds like a poor way to operate the system. We clearly need a better way to operate the system (see

Figure 6.2). Can anyone think of a smarter way to operate elevators in a 60-story building?

You could have all the elevators wait at the first floor. Elevator A could serve floors 1 to 20; Elevator B, 20 to 40; and Elevator C, 40 to 60.

This is certainly a better way to run the system. You could design the system in "banks," so that elevator A serves floors 1–20, elevator B serves floors 20–40, and elevator C serves floors 40–60. So, if you want to go to floor 53, you do not have to wait for those people that want to get off on floors 1 through 40. You simply get into Elevator C and go directly to the fortieth floor, and from there the elevator provides local service to the fifty-third floor. Who do you think would not like this service too much?

The person who is going from floor 53 to floor 22.

Why not?

People like to go directly. If you were going from 53 to 22 you would have to stop at 40, get on the 20 to 40 elevator, or perhaps go all the way down to the ground floor and go back up.

FIGURE 6.2 *Elevators with "banks."*

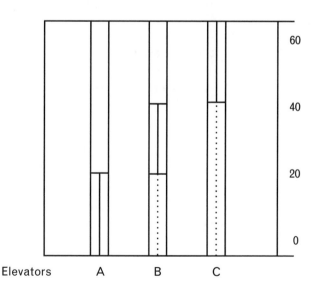

Good—this is an example of *redundant paths in the network*. Redundant paths provide more than one option between origin and destination, and the customer would make a judgment as to which path was better. We will discuss this later in greater detail.

Who else might not like this operation pattern? Consider a person who works on the fifteenth floor: What do they think of the system? Do they like it?

How much service does the person going to the fifteenth floor get, compared with what he used to get? He used to be able to walk in the lobby and take any elevator. Now he has only one elevator. Perhaps it cycles more often since it only goes to the twentieth floor, but he still only has one vehicle. Also, the fact that everybody getting into elevator A is going somewhere from floor 1 to floor 20 means that he is more likely to encounter people who want to get off before he does, because they are all 1 to 20-type customers.

The elevator configurations in Figures 6.1 and 6.2 represent rather different ways of using the vehicles, the three elevators. The time it takes each elevator to move through the building will be different in these two configurations. The number of passengers we can move and how quickly we move them will be different as well, as will the productivity of our assets, the three elevators. This illustrates the *vehicle cycle*, another key point in the analysis of transportation systems. All vehicles, be they elevators, freight cars, airplanes, buses, or ships are a fundamental and often expensive part of transportation systems. Keeping that asset productive is key to success.

Key Point 4:   The vehicle cycle

4. *Analyzing the flow of vehicles on transportation networks, and defining and measuring their cycle, is a basic element of transportation systems analysis.*

Next, we recognize that the level-of-service (LOS) provided is a function of volume. Whether all elevators go everywhere or they operate in banks, the more people that use the elevator, the poorer the service is in the aggregate. Why is that? We have already talked about one reason. There are probably some more. Yes?

More stops?

Just looking at it simply, if twice as many people want to use the elevator today as did yesterday, and I want to go near the top, I will probably not be alone in the elevator and my service will be poorer.

Do you think people may be more uncomfortable as volume goes up?

Comfort during the trip is another LOS variable. If you cram people in, the LOS goes down—less "hugs," as previously referred to. Also, you may have queuing—people may not be able to get on the elevator. For example, the elevator stops and the doors open but it is already packed and you cannot get on. So, in terms of service time, the level-of-service may deteriorate. Yes?

You also have increased times when you're stopped. The person who wants to get off is in the back of the elevator, so the elevator "dwells" longer.

Good. We will have more to say about the LOS/volume relationship later.

We just mentioned the fifth of the key points—the notion of queuing. In this particular instance, you wait for elevator service. Further, the system has to allow for storage. You have to have room for people to wait—that is, to queue. We have systems in which the vehicles wait or queue for passengers. Consider the taxi stand right outside of MIT on Massachusetts Ave. Taxis are often waiting for customers. Depending on time of day, the weather, and other factors, there are typically several taxis waiting. In this case we have a storage element for vehicles. Occasionally, I will go out there and there will be no cabs so I will have to queue for service. Queuing and storage for vehicles and customers is a recurring feature in transportation systems.

Key Point 5:   Queuing and storage
5. *Queuing for service and for customers and storage for vehicles/freight/ travelers are fundamental elements of transportation systems.*

Another point illustrated by the example is that transfers between elements of the transportation system are often inefficient. In the elevator example, a transfer from the walk-mode as one comes into the building, to the elevator-mode, implies some waiting and, hence, some inefficiency.

Key Point 6:   Transfers

6. Intermodal and intramodal transfers are key determinants of service quality and cost.

Another issue that comes out of this simple elevator example is operating policy, and the relationship of operating policy to LOS. We have already hinted about what operating policy means in this simple case. What is an example of operating policy?

The elevator would go up and wait until somebody requests it before going down.

That is one possibility. The question is where, and for how long the elevator dwells. What do you do with the elevator when it finishes dropping off passengers on the up-cycle? One operating strategy might be that when the elevator gets to the top, send it down, and pick up anybody who happens to need service on the way. Another operating strategy might be to wait up there until somebody requests service going down. There are all sorts of possibilities that have different level-of-service implications. Obviously, in more complex transportation systems, like an airline, operations is a much more complex issue.

Here's another example. Suppose I'm an early riser and I walk into my office building at seven o'clock in the morning. The place is empty and all three elevators are waiting in the lobby. I get into the elevator. I want to go to 42, and I get into Elevator C and press 42. The elevator sits there for a minute-and-a-half because it is programmed to wait, even though I know that it is unlikely that someone else is coming. Now, in principle, there might be some advantage to dwelling. If, by chance, somebody did come, they would have a better level-of-service. However, because I'm waiting for the elevator to move, my level-of-service is poorer than it would be if the elevator left immediately.

Key Point 7:   Operating policy

7. Operating policy affects level-of-service.

Another key point deals with transportation system capacity. Recognize that capacity is affected by many variables. We can add capacity to a transportation system by adding infrastructure. In this particular case, we could increase the number of elevators. We can also change vehicle technology. For example, we could have larger or faster elevators.

Additionally, we could have capacity improvements as a result of control technologies and smarter algorithms for dispatching. All this suggests our eighth key point.

Key Point 8:   Capacity

8. *Capacity is a complex system characteristic affected by: infrastructure, vehicles, technology, labor, institutional factors, operating policy, external factors (e.g., clean air, safety, regulation).*

Next we comment on one of the cornerstones of transportation systems analysis. This is the idea that the level-of-service provided by a transportation system is a function of the volume carried. We call this the *transportation supply relationship*. In many transportation systems, there is a close relationship between level-of-service provided and the fraction of system capacity that is being absorbed by volume. Specifically, level-of-service deteriorates dramatically as volume carried approaches system capacity.

Key Point 9:   Supply

9. *Level-of-service = f (volume); transportation supply. As volume approaches capacity, level-of-service deteriorates dramatically—the "hockey stick" phenomenon (see Figure 6.3).*

FIGURE 6.3   *LOS versus volume: the hockey stick.*

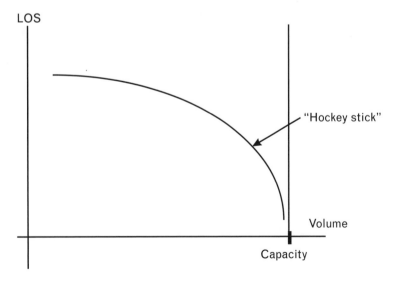

To illustrate the next key point, let's go back to the Volpe Center elevator example. What does the elevator control system know at any given time? It knows where the elevators are. It knows where demand for future service is and whether people want to go up or down. Further, it knows where the people currently in the elevator want to stop.

What doesn't the elevator control system know? Perhaps it does not know how close to capacity the elevator is. It also does not know what is likely to happen in the near-term future. For example, is there a big meeting breaking up on the twelfth floor, leading to a number of people wanting to go from floor 12 to the lobby? It does not know how many people are waiting at each floor and in particular, in the lobby. Perhaps it does not even know the time of day. If it did, it would recognize that at 5:00 p.m. a large peak loading is likely to occur as people leave at the end of their workday.

The basic idea is that with more information, the elevator system could be run more efficiently and effectively. The whole field of intelligent transportation systems (ITS) is based upon having real-time information about vehicles on highways and making network control and individual routing decisions based on that information. This is our next key point.

Key Point 10:   Availability of information

10. The availability of information (or the lack thereof) drives system operations and investment and customer choices.

An important question: can we make effective use of the information? For example, can we use the information to improve network control strategies and hence performance? Are there algorithms that we can utilize to make the network run more effectively? Can we perform those algorithms in the appropriate time frame? Specifically, can the algorithms be performed in real-time?

CHAPTER 7

Transportation Systems: Key Points 11–17

Key Points Continued

Let's go on to discuss our next key point: the "shape" of the transportation infrastructure has an effect on the fabric of the overall structure. The overall structure goes beyond physical structures, like buildings, to the structure of an urban area or even a nation.

In the elevator example, we have made a design decision to have three elevators that serve the floors as described. Now, let us imagine we are back in the design phase and not constrained to three elevators (see Figure 7.1). Some genius could have said, "What we really need is direct service to every floor. So, what I suggest is building 60 elevators

FIGURE 7.1
Direct elevator service.

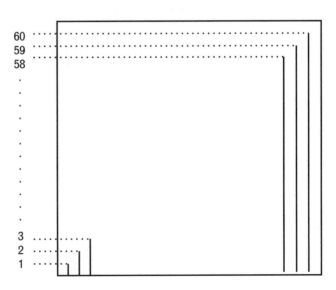

in this building. There will be one that goes from floor 1 to floor 2, and one that goes from 1 to 3, and one that goes from 1 to 4, and one that goes from 1 to 5; everybody will have direct service."

Now, it is obviously a foolish idea but, in a sense, it is effectively addressing the level-of-service experienced by the customer. Direct service creates a terrific level-of-service. Unfortunately, the investment in infrastructure is so extraordinary and so modifies the fundamental shape of what you are trying to do—in this case, provide an office building—that it becomes ridiculous. If we actually had to build 60 elevator shafts, we would not have any room left for offices. The office building would be 60 elevator shafts, and you go up, find nowhere to go to, and you come back down. But, as in Figure 7.2, would it be ridiculous to build 6 shafts with services to 10 floors each?

The configuration in Figure 7.2 may make sense, depending on your perspective. You should think about the costs and benefits of having some number of elevators.

Now, some urban planners might argue that the 60-shaft solution or even the 6-shaft solution is not unlike our current urban highways. Some argue that we have chosen to develop urban highways in a way that is as bizarre as providing 60 elevators for the 60-floor building. By providing the universal, "go from anywhere to anywhere, anytime you want" highway system in the urban area, we have provided a transportation system that has stretched the urban fabric in such a way that its whole purpose is warped. Some argue that by building vast amounts

FIGURE 7.2 *Another elevator configuration.*

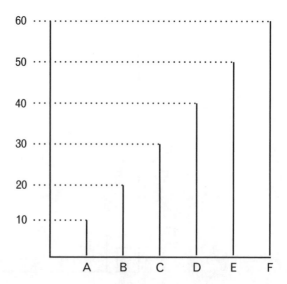

of highway infrastructure to provide transportation service, we have exceeded rational investment and have diminished the functionality and livability of our cities.

Key Point 11: Infrastructure shape

11. The shape of transportation infrastructure impacts the fabric of geo-economic structures.

For our next point, we must recognize that different customers get different levels-of-service, depending on system configuration and operating policy. The idea here is that service is not provided uniformly to all customers or, in the elevator example, all people that work in the building. People get different levels-of-service, depending upon what floor they are on and how the system is operated.

Again, to understand this, we can think of various real-world transportation systems where level-of-service provided is, in fact, quite different for different customers using the same system. In a bus network, the routes affect who gets better and worse service.

If the urban bus system charges the same fare for all customers who experience different levels-of-service, there may be inequities. In thinking about transportation systems design, we must consider who benefits and who pays for the system. This can be a rather subtle question, and often fairness has very little to do with it. This leads us to the next key point.

Key Point 12: Costs, prices, and LOS

12. The cost of providing a specific service, the price charged for that service, and the level-of-service provided may not be consistent.

Further, our ability to compute costs for serving particular customers and for providing particular services is often limited. For example, if a passenger train and a freight train both use the same rail line, how do we allocate the maintenance costs for that line to passenger and freight service? This suggests the following key point.

Key Point 13: Cost of service

13. The computation of cost for providing specific services is complex and often ambiguous.

The next key point deals with the idea of cost/LOS trade-offs. As introduced earlier, we talked about the vice-president of operations, and the vice-president of marketing for a railroad. Arguing about cost and service, the VP-Operations says, "Gee, we can't run trains that frequently—it costs too much," and the VP-Marketing says, "If you don't run those trains, we won't get any traffic, and what's the point of running trains at all if we do not have business?" We will see this trade-off many times.

How might this cost/LOS trade-off be illustrated in the elevator example? Think about the idea of dispatching policy. Consider the floor 1 through floor 20 elevator. You could run this system such that when the elevator comes down to the first floor, it is dispatched two minutes later on another cycle through the building. There would be a level-of-service associated with that particular operating policy. Another idea is to dispatch the elevator when it is full. Suppose an electric eye counts people when they enter: when 12 people are on the elevator, up the elevator goes. Is this a good system?

> At the a.m. rush hour, the elevator fills up rather quickly; it is probably not bad. At off-peak, from a service point of view, it might be atrocious. It would be frustrating to stand there with 11 people in the elevator waiting for a twelfth. You might be paying people to take the elevator with you so that it would leave.

Another way of operating the elevator is not that different from the way some railroads operate: there could be a trade-off between waiting time and load relative to capacity. For example, you could have a dispatching policy like the one illustrated in Figure 7.3.

In this example, you would dispatch the elevator immediately with 12 people aboard. So, if it is the morning rush hour and everyone is going to work, when the elevator comes down to the first floor, at least 12 people get on, and off goes the elevator. (Note that the capacity of the elevator could be greater than 12.)

Now, suppose there is no one in the elevator at time zero, and people start to enter. When the cumulative number of people in the elevator reaches some number, the elevator is dispatched, as shown in Figure 7.4. Basically, we do not want our customers to wait a long time for the elevator to leave (affecting their LOS). At the same time, we do not want to send an elevator up for every person, which would be very inefficient from an elevator-cycle viewpoint.

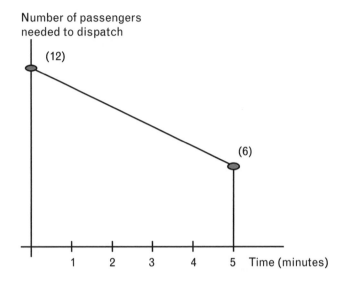

This is similar to a policy railroads sometimes use for building a train. Their concern is the length of the train. With trains coming in from stations west (Figure 7.5) they want to build an outbound train that will have cars from many different inbound trains. When do they dispatch the train from the terminal? They are balancing the LOS for the cars already on the outbound with those that have not yet arrived. If they dispatch the train, the latter cars will need to wait for a subsequent outbound train.

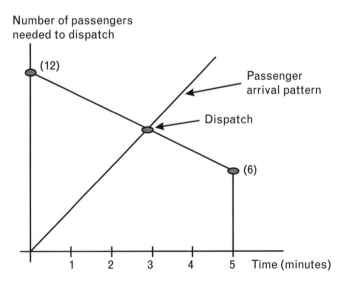

FIGURE 7.5
Building a train.

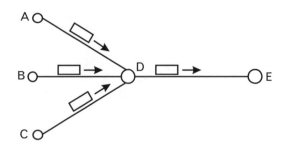

So there are many ways we can operate the system—each has a service and a cost attribute. The operating cost of waiting until the elevator is full is low. We would get a very good load factor, but the elevator is not dispatched often so the level-of-service is likely to be poor.

On the other hand, the system that dispatched elevators too often may be expensive to operate (e. g., power requirements, wear-and-tear, and so forth). The level-of-service may be very good, but the costs may be high. Actually, if we dispatch too often—say one elevator per customer—the level-of-service may be poor since the elevators are spending a lot of time running empty (or with only one customer) and people would have to queue for service in the lobby. This all illustrates our next key point.

Key Point 14: **Cost/level-of-service trade-offs**

14. *Cost/level-of-service trade-offs are a fundamental tension for the transportation provider and for the transportation customer, as well as between them.*

Costs are always a concern in transportation systems operation. Often we will see that batch processing, or the consolidation of like-demands, is used as a cost-minimization strategy. The previous railroad example is a good one. In the elevator context, the batches would be the customers who want service from floors 1–20, 20–40, and 40–60. Of course, as always, the cost-reduction strategy of consolidation has service implications.

The concept of consolidating traffic will come up again and again in transportation systems. When an airline runs a hub-and-spoke operation, it is consolidating people from different origins who have common destinations into airplanes to lower costs. Consolidation happens in rail freight systems, as is done in many kinds of transportation systems.

Key Point 15: Demand consolidation
15. Consolidation of like-demands is often used as a cost-minimizing strategy.

Next, we talk about investments in capacity. Operators can invest in capacity in many different ways. These investments in capacity are lumpy—a step-function—as opposed to continuous. For example, if we have two elevators as opposed to three elevators, the performance of the elevator system changes quite a bit. One of the inherent difficulties in transportation system design is that the investment is lumpy—we cannot buy two-and-a-quarter elevators—we have two elevators, or we have three elevators. So, we end up with a function like Figure 7.6.

We have a lumpy function for capacity and a lumpy function with respect to investment. Another example is a rail system. One alternative is to build a single track between two points on our network. But, since we want to provide service in both directions, we need to provide some mechanism for trains to get by each other—these are not trucks; they have to stay on the track. We build what are called sidings, which would permit us to pull a train off the mainline, so a train coming in the other direction could get by. Sidings also allow a faster train going in the same direction to pass.

Another alternative is to build a piece of rail infrastructure with double tracks between origin and destination (see Figure 7.7). We would have eastbound service and westbound service with all the trains

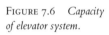

FIGURE 7.6   *Capacity of elevator system.*

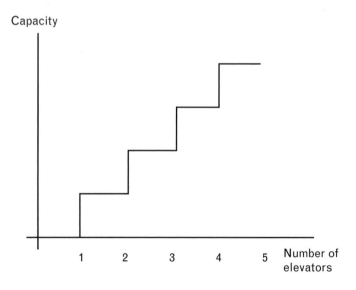

F<small>IGURE</small> 7.7   *Capacity of single versus double track rail line.*

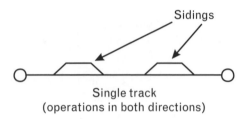

Single track
(operations in both directions)

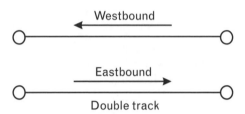

Double track

going eastbound on one track and all the trains going westbound on another, with no interference between the two.

So, we are making a judgment about investing in infrastructure—constructing two tracks is going to be more expensive than building one with a few sidings. We have made some judgments about cost and capacity and level-of-service in a "lumpy" way.

Key Point 16:   **Lumpy investment**
16. Investments in capacity are often lumpy (e.g., infrastructure).

Now, consider our three-elevator building. If we construct our building with three elevators that go from the first floor to the sixtieth floor (see Figure 6.1), we have made an infrastructure investment. We have the operating option—even though the infrastructure exists to go to the 60th floor for all three elevators—of operating elevator A from floors 1 to 20, elevator B from floors 20 to 40, and elevator C from floors 40 to 60.

Now, we have three 60-floor shafts. We are not taking advantage of it at the moment, but we have flexibility. If the patterns of usage on these elevators change, we have the ability to change our operating pattern. So, for example, if it turned out that we have a lot of demand on the upper floors, we might change the way in which we operate the elevators (e.g., elevator A to floors 1–40, elevators B and C to floors 40–60). However, if we choose to build the building less expensively, with the elevator shafts going only to floors 20, 40, and 60, respectively—the shafts are not physically there—we would not have that option (see Figure 7.8). We pay for operating flexibility.

One of the problems with the simplicity of our elevator example is that by virtue of the system configuration, we are talking about one elevator, one shaft. (It is not easy to operate more than one elevator in one shaft—they do not go by each other very well.) In many other systems, such as urban bus systems operating on highways, the vehicles do, of course, pass each other. In bus systems we can add capacity in a much less lumpy way than by adding elevator shafts. We can add capacity to an urban bus system in a relatively continuous way by simply buying more buses (see Figure 7.9).

The same is true in a rail system. Here we can add freight cars. There are several million freight cars in this country, so clearly there is a fairly continuous relationship if one wants to add capacity by adding vehicles rather than by adding infrastructure, which is what we are doing with elevator shafts.

However, there are intermediate options when investing in capacity. There are several orders of magnitude fewer locomotives in the United States than there are freight cars. So, there is a range of "lumpiness." If we construct a rail line, we choose between single or double tracks. If we add capacity by adding locomotives, it is less lumpy than infrastructure. But it is still lumpy. If we add capacity with more freight

FIGURE 7.8 *Another elevator configuration.*

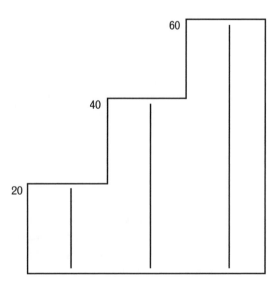

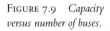

FIGURE 7.9   *Capacity versus number of buses.*

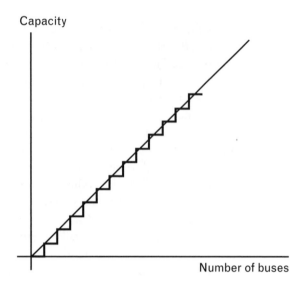

cars, the investment is essentially continuous, although, in theory, it is still lumpy.

What other kinds of resources can we talk about? We have talked earlier about vehicles, infrastructure, and control systems. What other kind of resources might we want to add to a system, or delete from the system for that matter, to change its capacity? In addition to those previously discussed, are there other kinds of resources that could, in effect, change the capacity of our system?

The number of service personnel at the airlines affects capacity.

Good. Labor is a resource that is related to capacity. In this particular case, we are saying we could have one person taking the tickets at the podium, and another making sure everybody has a boarding pass, as opposed to one person doing both. That might, or might not, have an effect on capacity. It probably does affect capacity, but not much. It would have an effect on level-of-service, certainly, or the speed with which passengers board the plane.

Here is another example—consider switching crews at railroad yards. If we add crews, we are, in effect, making a lumpy investment in labor—the throughput of the railroad yard would change. So we see that we have a variety of ways to affect capacity, as previously introduced in Key Point 8.

How about clean air? Is that a resource that we have to think about as a transportation decision-maker?

Well, in Southern California, most transportation decisions right now are being made or justified in terms of the Clean Air Act, not in terms of meeting any idea of customer service or what people want.

So we have to think of clean air as a resource that affects capacity, just like a locomotive is a resource or your automobile is a resource.

Think of resources and think of investments in capacity in a very broad way. (Refer back to Key Point 2: Transportation—part of a broader system.)

There are obvious resources like infrastructure, vehicles, labor, and control systems, but there are other resources that we use. Perhaps by making clever use of particular kinds of technologies, we can expand the capacity of our system by, say, using less clean air through the use of electric cars, which is one concept being advanced [1].

Lumpy Elevator Investment

In the elevator example, the decision about the number of elevators was lumpy—it is not possible to have two-and-a-half elevators. If we plotted the capacity of the system on the Y axis, and the number of elevators on the X axis, we would see a step function.

Extending this discussion, there is a relationship between *cost* and investment in capacity—be it in infrastructure, vehicles, control systems, etc. As we buy capacity, we are purchasing it in a lumpy fashion as well. If we purchase another elevator or a few more buses, it not only changes the capacity, but it changes our cost structure as well. That is, the fixed cost of operating the system is more substantial, and these costs are lumpy (see Figure 7.10).

There is a relationship between the cost and capacity of a transportation system. However, recognize that there is a trade-off. We can have inexpensive, low-capacity systems or more expensive, high-capacity systems. Making decisions about which systems we need or want is a continuing theme in transportation.

Additionally, we should note that the concept of level-of-service is related to the cost and capacity trade-off. That is, the level-of-service provided by the transportation system is a function of the volume being

FIGURE 7.10   *Capacity versus cost.*

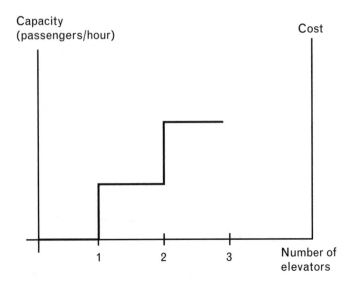

carried by that transportation system (see Key Point 9—Transportation supply).

For example, if we have a high volume of people that want to use the elevator, the elevators will be crowded. We may have to queue and wait for a few elevators to come before it is our turn. The level-of-service, as measured by comfort and by elapsed travel time, will deteriorate. The typical relationship between level-of-service and volume is shown in Figure 7.11.

The level-of-service deteriorates as the volume approaches full capacity. Often, there is a precipitous fall-off in level-of-service as the volume approaches capacity, sometimes called a "hockey stick." (Also, see Figure 6.3.) So, we have to make decisions about how much we want to invest in our transportation system, given lumpy investment patterns and the hockey stick phenomenon that often characterizes the relationship between volume and level-of-service as the volume approaches full capacity.

This leads to one of the most challenging aspects of transportation system design. If we underinvest in capacity, our level-of-service may be uncompetitive. If we overinvest, level-of-service may be fine, but costs will be high and our prices may not be competitive. Making this decision, faced with lumpy investments and the hockey stick LOS/volume relation, is difficult indeed.

FIGURE 7.11   *Typical supply functions.*

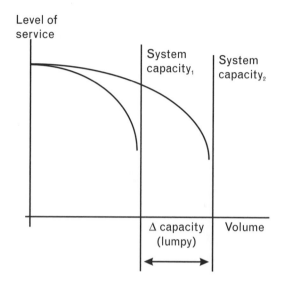

Key Point 17:   Capacity, cost, and level-of-service

17. *The linkages between capacity, cost, and level-of-service—the lumpiness of investment juxtaposed with the hockey stick level-of-service function as volume approaches capacity—is the central challenge of transportation systems design.*

REFERENCE

1.   Sperling, D., *Future Drive: Electric Vehicles and Sustainable Transportation,* Washington, DC: Island Press, 1991.

CHAPTER 8

Transportation Systems: Key Points 18–24

....
Key Points Continued

Let us introduce our next concept—the concept of "peaking" in transportation systems. Consider our elevator example. We plot the kind of volume that we would expect to see on the elevator system as a function of time of day.

We start by looking at the up direction. Assuming that people are using the elevator to get to work and that most in this building work 9:00 a.m. to 5:00 p.m., we could suggest a relationship between time and volume that this system carries over a typical day, as shown in Figure 8.1. This is the kind of phenomenon that we see very often in transportation systems—sharp peaks in demand over a day. This kind of pattern is associated with journey to and from work.

A key question that we, as the operators of the system, must ask is how much capacity to provide, recognizing that we do not have a constant demand over time. We could make the judgment that we are going to provide for the maximum demand as in $Capacity_1$ (see Figure 8.2). This system will operate so that the volume never exceeds the capacity. Indeed, we will have a high LOS system, but at the same time, of course, we will have high-cost system, because we have made an investment in capacity that, while it may look pretty sensible at 9:00 a.m., doesn't look quite so smart at 1:30 in the afternoon.

On the other hand, we could say on the average, our hourly demand is "X" people per hour and we want to build a system that provides for the average demand as in $Capacity_2$. We choose to ignore the fact that demand is temporally distributed. On the "average," the

FIGURE 8.1   *Volume versus time of day.*

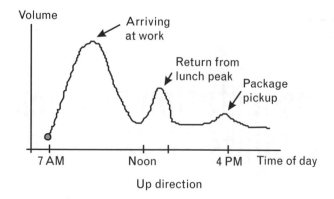

Up direction

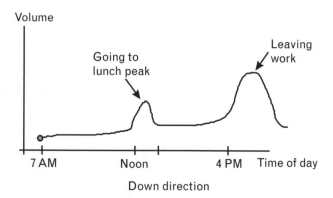

Down direction

FIGURE 8.2   *Different capacity decisions.*

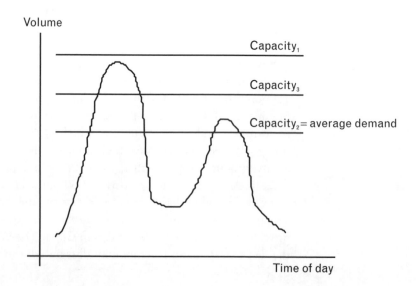

system works. When the volume exceeds capacity, as it will during peak periods, customers will have to wait for service—and service will deteriorate. Despite some customers having to wait for service during the morning peak, all customers will eventually be served and the system will recover after some period of time.

So in this case, we have a system that is much lower in cost from an infrastructure point of view. But as a result of this peaking relationship—and peaking is characteristic of transportation demand—a lot of customers, specifically those who come during a peak period, will receive poor LOS. There will be queues; people will have to wait before they can get on the system.

So what do we do? We cannot choose such a low capacity that customer levels-of-service during peak periods are unacceptable. At the same time, however, we cannot provide a level-of-service such that nobody ever has to wait—it's not economical. So, Capacity$_3$ may be a good compromise. The question of design capacity and how we accommodate temporal peaks in demand is Key Point 18.

Key Point 18:	Peaking

18. Temporal peaking in demand: a fundamental issue is design capacity—how often do we not satisfy demand?

Now, suppose the elevator system has capacity so that if you arrive at 8:55 a.m., you do not arrive at your office on the forty-second floor until 10:30 a.m. After three weeks of this (and your boss giving you fishy looks as you walk in at 10:30) you might say, "I really want to keep this job. I need this job. Instead of getting to the lobby at 8:55 a.m. and waiting for an hour and 25 minutes, maybe I'd better get there at 8:20 and wait only a half-hour. Maybe then I'd get to work on time."

So depending on the circumstances, you might travel outside the peak—either before it or after it. If you could increase your LOS, you would be enticed to move yourself off the peak. We will come back to this point later.

In Los Angeles there is broadening of the peak-hour traffic. People who want to get to work in time leave for work around 6:00 a.m.

This happens outside of California, too. In cities like Bangkok, where congestion is truly atrocious, peak hour extends virtually throughout the day. The system is running close to capacity virtually all the time.

The next major point we must recognize is that the volume that a transportation service attracts is a function of the level-of-service provided to customers. If the level-of-service deteriorates, fewer people will want to use the service. How many fewer, of course, varies from case to case. The idea, however, is that the volume attracted is a function of the level-of-service provided. This is simply a micro-economic concept. For example, if a movie theater doubles its price, therefore making its service less attractive—in this case, more expensive—fewer people will go to that movie theater. If a movie theater halves its price, more people will go.

A relationship between demand and the provided level-of-service is shown in Figure 8.3. In general, we assume that people like to buy inexpensive goods. So if, for example, we are providing cheap (i.e., fast, safe, reliable) highway transportation service, people will tend to "buy" more of it than if we are providing expensive (i.e., slow, less safe, unreliable) transportation service. This is a fundamental economic argument and one that is central to demand for transportation services.

Key Point 19:   Transportation demand

19. Volume = f(level-of-service); transportation demand.

We have talked before about the idea of level-of-service, as a general, multidimensional concept. We simply cannot measure it by a single variable—like travel time. Rather, we measure it with a number of level-of-service variables. Often, we have to collapse several level-of-service variables into a single variable—utility. For example, utility V might be the linear sum of various LOS variables, such as travel time, access time, waiting time, fare, and comfort, as described in Chapter 4. Other functional forms may be used as well.

FIGURE 8.3 *Transportation demand: LOS versus volume.*

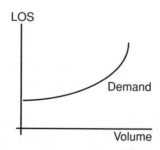

Key Point 20:   Multidimensional level-of-service

20. Level-of-service is usually multidimensional. For analysis purposes, we often need to reduce it to a single dimension, which we call utility.

In the elevator example, suppose the building is in place and that some particular level-of-service is being provided. Then a new tenant moves in on the forty-seventh floor, with another 200 employees, and the volume that the elevator is required to carry goes up. With 200 new employees, the elevator service is a lot worse. What do you suppose would happen in the short run, over the next week or so?

People would start walking up the stairs rather than using the elevator.

That is a possibility. But, I would argue that in a skyscraper, your modal choices are rather limited. Maybe it works to walk up for the first three floors. And maybe for the real athletes, it works for the first eight floors. But, assuming that people still have to go to work, the volume, at least in the short run, will not change in any substantial way. Volume might change in the downward direction. People are more inclined, for obvious reasons, to walk down rather than up. So people might walk down from the eighth floor, where walking up to the eighth floor is unlikely. In general, the modal opportunities are limited—there are no competing modes—and assuming that people working in this building want to keep their jobs, the volumes will not change much.

However, in situations in which there is modal choice, we would see a decrease in volume resulting from a deterioration in service. So, for example, you are driving to work by car every day, and traffic is getting worse and worse. Finally, one day you throw up your hands and say, "I'm going to put this car in the garage and I'm going to take the commuter rail," and you shift.

Indeed, this is probably true in our building with poor elevator service, too—it will just take a while longer. What would happen in the building?

Eventually, people would move out of the building.

Right. General Electric, who has a sales office on the forty-second floor, eventually says, "This is ridiculous. I'm paying a premium rent for this space, and it takes my people 20 minutes to get up here every morning. There's a new building opening down the street and I'm going to move because the elevator level-of-service here is so poor." And, indeed, the elevator volume demanded in our building will go down. It will not happen tomorrow. It is going to take a while, but in the long run it will happen.

As noted in Key Point 2, competition is the underlying driver of improved transportation service. In the case of the elevator system, there is little modal competition—if tenants stay in the building, they are captive to the elevator. More usually, there is competition. For example, if you raise rates for the trucking service you provide, or if LOS deteriorates in other ways, your customers will switch to a competitor—say, rail or another trucking company. In the elevator case, competition occurs at another level—people eventually move into a different building altogether. Competition exists in a different timeframe—strategic rather than tactical.

Manheim discusses this changing of the activity system in response to the transportation system (and vice-versa) in his conceptual framework.[1] In using the framework, we have to distinguish between short-term, medium-term, and long-term effects. The activity system changes over the long term, typically.

Key Point 21:   Different time scales
21. Different transportation system components and relevant external systems operate and change at different time scales (e.g., short run—operating policy; medium run—auto ownership; long run—infrastructure, land-use).

Now, we introduce equilibrium between supply and demand. We have noted that, on one hand, level-of-service is a function of volume. As a facility gets congested, the level-of-service for customers deteriorates, as shown in Figure 6.3 and as discussed in Key Point 9, Transportation supply.

On the other hand, looking at it from a micro-economic perspective, as level-of-service changes, demand changes as well. Demand increases as level-of-service improves and decreases as level-of-service deteriorates.

We show this in Figure 8.4. Here we have a demand function that shows a relationship between level-of-service measured in some

---

1   Manheim, M. L., *Fundamentals of Transportation Systems Analysis*, Vol. 1, Cambridge, MA: The MIT Press, 1979. This book is a classic treatment of transportation systems analysis.

FIGURE 8.4
Equilibrium.

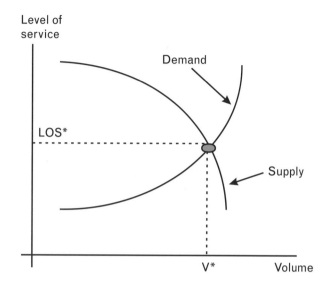

unidimensional way—using the utility concept—plotted against volume. However, there is some maximum volume, regardless of the level-of-service provided. We also have a supply function, as introduced earlier, which basically says that as volume increases, the level-of-service provided to customers decreases, and vice-versa.

The fundamental idea is an equilibrium between supply and demand, so that the system operates at some equilibrium volume V^*, and at some equilibrium level-of-service LOS^*.

Key Point 22:   Equilibrium

22. *Equilibration of transportation supply and demand for transportation service to predict volume is a fundamental network analysis methodology.*

This is a useful way of thinking about transportation systems. We should emphasize, however, that it is conceptual. It is a framework for thinking about transportation that may not be completely correct in detail but is nonetheless a useful way of conceptualizing transportation systems.

This simple framework overlooks, for example, that we have different users with different sensitivities to level-of-service variables. It also overlooks the fact that we have different time-scales involved in different parts of the system (Key Point 20). For example, an individual customer can switch from highway to transit overnight. To build highway lanes takes a long time. So, a system may tend toward equilibrium but never actually reach it or may reach it only after many years. Thinking about our elevator example, time-scales on both the supply and demand sides vary (see Table 8.1).

This simple framework overlooks the competitive aspect of transportation. If you, as a trucking company or as a railroad company are changing your level-of-service, your competition is not going to sit there and watch you do it (Key Point 3). There may be imperfect information about the level-of-service provided to the users (Key Point 10).

So, for any number of reasons, this supply/demand equilibrium model is not totally correct. However, I would argue it is still a very useful way to think about transportation systems. A way to capture this thought—the words of an anonymous sage—is as follows:

All models are wrong.
However, some are useful.

Any model will, at some level of abstraction, fail to capture some aspect of the real-world system. Models can still be useful representations for the purposes of understanding the system, for the purposes of designing the system, for the purposes of changing operating plans, and so forth. So indeed, any modeling effort—any reduction to an abstract form of a system is, in some sense, wrong. However, it may still be very useful. Certainly, many of the models we will talk about are both wrong and useful.

Soon we will talk in more detail about modeling issues and how they relate to transportation systems (see Chapter 10, "Models and Frameworks," and Chapter 11, "Modeling Concepts"). For now:

TABLE 8.1   Demand and Supply

	DEMAND	SUPPLY
Short run	Some people walk up and down	Elevator operations—change dwell times
Medium run	Firms change work hours	Technological fix—new control system
Long run	Firms move out	High-tech, high-speed elevator More shafts

Transportation systems are complex, dynamic, and internally interconnected as well as interconnected with other complex dynamic systems (e.g., the environment, the economy).

They vary in space and time (at different time scales for different components). Service is provided on complex networks. The systems are stochastic in nature.

Human decision-makers with complex decision calculi make choices that shape the transportation system.

Modeling the entire system is almost inconceivable. Our challenge is to choose relevant subsystems and model them appropriately for the intended purpose, mindfully reflecting the boundary effects of the unmodeled components.

The Mechanics of Supply/Demand Equilibrium

Let us now discuss some of the mechanics inherent in supply/demand equilibrium. Often, we need to use some kind of an iterative numerical procedure to find the equilibrium point where the supply and demand functions intersect. It is sometimes straightforward—but often it is not.

In supply/demand equilibrium, we choose a volume that we guess is close to the equilibrium point. Based on that volume, we compute a level-of-service (that level-of-service will be inconsistent with the volume we selected initially). We then change the volume to be consistent with that level-of-service, and then the volume is incorrect. So, we go through an iterative process that allows us to compute the equilibrium point (see Figure 8.5).

Changing Supply

Now that we understand the model, let's talk about using it. Suppose we have measured level-of-service and volume on the system. Further, we have a demand relationship and a supply relationship. Let us also suppose that this is a highway system and we have decided that we want to add to the infrastructure base, and presumably improve our supply of transportation services. This can be conceptualized by showing the supply function, moving in this case, from left to right (see Figure 8.6).

FIGURE 8.5   *Iteration to find equilibrium.*

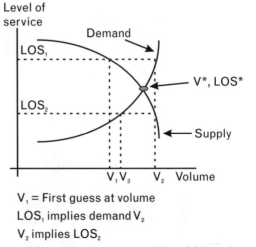

V_1 = First guess at volume
LOS_1 implies demand V_2
V_2 implies LOS_2
LOS_2 implies V_3, etc., until V^*, LOS^* are found

So, while we had an equilibrium with a level-of-service LOS* and a volume V*, we have now added to the supply of the system. We would expect that over some period of time, the system would adjust to a higher volume V** and a new level-of-service LOS**. We see volume grow to accommodate the fact that we are, in effect, making highway transportation cheaper (i.e., faster).

FIGURE 8.6   *Changing supply.*

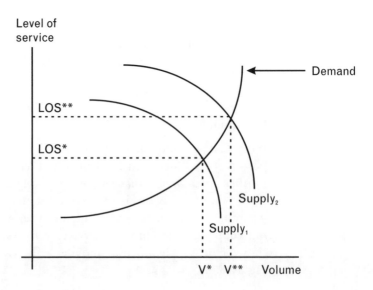

People respond to incentives. If highway
transportation becomes cheaper, people buy more.

Changing Demand

Now, where are these people coming from? Are they moving from
transit, for example, to highway? Are they moving from the turnpike to
this highway? There are all sorts of explanations for where that demand
might come from. There may be latent demand—people who did not
travel at all before, but now choose to travel because it's cheaper.

What might happen if Route 2 doubled in capacity as we improved
it over a period of time (see Figure 8.7)? Before we improved the highway, it used to be difficult to get from Fresh Pond to Route 128. Now
it is easier, because Route 2 is a wider, higher-quality road. Along
comes a developer and says, "Let me put my Raytheon headquarters at
the intersection of Route 128 and Route 2." On the other side of
Route 2, some real estate developer says, "Let's build some new
condominiums."

As a result of the better level-of-service provided by the expanded
Route 2, economic activity is attracted to the area. Over some
period of time, the demand curve shifts from **Demand₁** to **Demand₂**
(see Figure 8.8).

We begin to see higher demand as a result of construction of the
highway. We have promoted activity by changing the infrastructure in

FIGURE 8.7   *Boston and*
west.

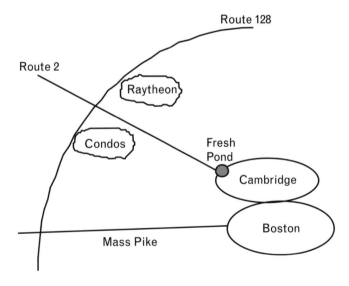

FIGURE 8.8   *Changing demand.*

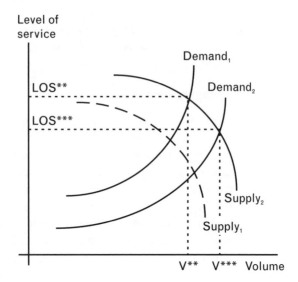

the area. So now, we are operating at yet another equilibrium (V***, LOS***) after some period of time. Over a long period of time, activity on the system adjusts, economic activity increases, new buildings are built, the demand for transportation increases, and we have a new equilibrium.

FIGURE 8.9
*Equilibrium:
a second look.*

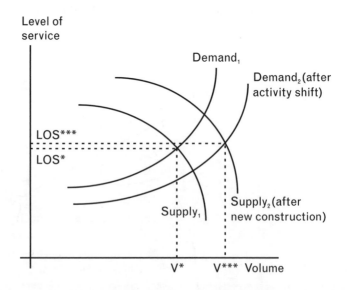

Looking at Figure 8.9, the level-of-service (LOS***) provided by the improved facility looks nominally equal to the level-of-service (LOS*) before we started. And this is the kind of story we hear again and again when we study transportation. "Gee, we had a highway, and it was congested—they expanded it and it's still congested."

Increased Volume: Better or Worse?

So, at a higher volume, the customers are seeing the same level-of-service. Are we worse or better off than we were before we started? Which would you prefer?

One could argue that with greater volume, there is more potential for environmental destruction than before.

"Destruction" is kind of an inflammatory word, but it is still a good point. There is certainly an environmental impact from this additional highway volume.

However, we could argue that volume is good. Volume indicates economic activity and people going to work. New volume could be lumber going to build new homes or people going to the movies to enjoy themselves. If volume is viewed as economic activity and there is now more of it than there was before, perhaps we are better off.

We could argue quite the opposite: volume has major environmental impacts. By increasing the volume, we are using up other kinds of resources that are not explicitly in this framework. We have talked about clean air as a resource that we are using. Maybe we are using up too much of it. Because clean air is not explicitly a cost to individual users, the pricing system is working in such a way that we are using up clean air at an inappropriate rate.

These arguments highlight the question of transportation sustainability: the concept is to balance economic development and environmental impact. Those who feel that volume is good are coming down on the side of economic development. Those who are more skeptical of increased volume are concerned about environmental impacts and the use of resources. There is no easy answer to the sustainability question. This equilibrium framework may be a good way of thinking about it.

....
Peaking

Let us return to peaking. You will remember we talked about daily peaking. Daily peaking, however, is not the only kind of peaking that one sees in transportation systems. There are other time frames in which transportation demand peaks, both in traveler and in freight modes. For example, weekly peaks are observable in many transportation systems. Highway traffic on Friday afternoons tends to be worse than on other days. Tuesday tends to be the lowest day of the week for air transportation in this country. If you have flown on Friday evenings, you know it is very different than flying on Saturday afternoon. We also have seasonal peaks in transportation systems. For example, in air passenger service, Thanksgiving is the largest peak of the year. In freight systems, we will see peaking at harvest time.

Returning to the elevator example, the elevators peak daily. Suppose the building population grows and more companies decide to locate in the building. All of a sudden, the elevators that used to run smoothly—we could walk into the lobby at 8:55 a.m. and be assured we would be in our office by 8:58—are overcrowded and we have to queue. There is congestion and delays at peak hours.

We discussed what people would do in this circumstance. We commented that some people would come to work early, and some people would come to work late, and people would try to figure out what to do to accommodate the peak. We might see the employers in the building decide to band together to come up with some ways of reducing peak service demands. Various renters in the building might try to stagger work hours, so that the peak was somewhat ameliorated. If the renters could agree as to which employer took which work hours, that would be a mechanism for reducing the peak.

Price as an Incentive

Alternatively, we could do what is often done in various systems. We could use price as an incentive to urge people off the peak. For example, we could charge for the use of the elevator. Suppose there is now a fare to ride the elevator. It used to be free, and elevator customers didn't have to pay; now they are going be charged. Forgetting about implementation details—which, in practice, are difficult—we could put in a system where, if customers wanted to take the elevator from 8:45 to 9:00 in the morning, it would cost X dollars.

If customers were willing to use the elevator say, between 8:30 and 8:45 a.m., or between 9:00 and 9:15 a.m.—immediately off the peak—we would let them ride for free. If they are willing to use the elevator before 8:30 or after 9:15 a.m.—really off the peak—we might pay them to use the elevator. (Now, obviously, there would have to be a system in place so people didn't come in off the street to take the elevator off-peak and receive payment.)

In principle, we could make that work. We could reduce peak demand based on the economic theory that people respond to incentives (see Figure 8.10). Pricing would be a mechanism for possibly accomplishing a peak profile that is more dispersed.

Congestion Pricing

This is called "congestion pricing." People who are willing to pay to travel in the peak would do so, and people who were not willing to pay would travel outside the peak period. Now, of course, it is not that easy, as we in transportation have learned. Transportation professionals have talked a lot about this concept, particularly in highway transportation. It is only recently that technology has made it reasonably possible to implement congestion pricing on highways.

With congestion pricing, there are a number of issues to consider. First of all there is the political issue of charging people to use something that was once "free." In the elevator case, there would be some money

FIGURE 8.10 *Changing the demand profile.*

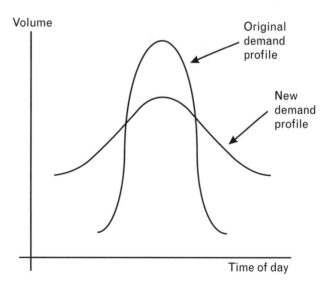

transfers taking place between people who were willing to pay the peak-hour price between 8:45 and 9:00 a.m. and people who were willing to travel either later or earlier, and would receive money.

Another issue, again political in nature, is the question of equity. The system of congestion pricing could be argued as favoring the rich. Not everybody can afford that $X a day. It also favors the flexible. There are people who work for their fathers-in-law, and they will come in whenever they like and, in doing so, also earn $Y a day because they're going to use the system at 11:00 in the morning, instead of at 9:00 in the morning. Then there are people whose bosses say, "I do not care what they are charging you to get on the elevator—be here at 9:00 a.m." There are equity issues that we should be concerned with in implementing congestion pricing schemes. Often, the people with the flexibility are the people with the resources to pay.

Congestion pricing is being considered in serious fashion in real-world transportation systems. With the advent of intelligent transportation systems (ITS), there are technologies that allow the collection of highway fees as a function of time of day or other conditions, without the car having to stop. This opens the possibility of congestion pricing schemes for highways. So, the idea of ameliorating peaks, using price, is indeed very much on the table.

Pricing of Transportation Services

We can generalize this point to the pricing of transportation services. In discussing Key Point 12, we noted that the price someone pays for transportation service might not be consistent with the cost of providing that service. Further, the costs imposed by our use of the transportation service may be imposed not only on us, but on others as well. For example, when we choose to travel at peak-hour, we cause further delays for our fellow travelers. When we drive, we impose costs—due to environmental degradation—on society as a whole, for which we don't directly pay.

Economists use the term "negative externality" for these effects. Our decision to use the system creates costs for others and may lead to suboptimal use of resources from a societal perspective. Pricing—like congestion pricing or charging people for the congestion they are causing—is a possible way to reduce negative externalities. This can be stated as Key Point 23.

Key Point 23:   Pricing

23. Pricing of transportation services to entice different behavior is a mechanism for lowering the negative externalities caused by transportation users on other transportation users and society-at-large.

Flow Imbalance

A characteristic of transportation systems is imbalance in flows. In our elevator system, we have an imbalance. In the morning, everybody wants to go up. In the evening, everybody wants to go down. In freight transportation, we have imbalances. There is a large amount of freight traffic flowing from Pacific Rim countries by ship to West Coast ports—Long Beach, Los Angeles, Seattle-Tacoma—and then via double-stack container trains to the Midwest and the East Coast. These containers are all going full from the West Coast easterly. However, those containers have to be transported empty to the Pacific Rim to be loaded again.

This is a fundamental element of many transportation systems. Flows are not balanced. We deal with it in the railroad industry—repositioning of empty cars is an example. We deal with it in the trucking industry with repositioning of trucks. The term used is "back-haul." Truckers are very anxious not to ride around empty. When truckers have a shipment from Schenectady to Pittsburgh, they try to find some other shipment in Pittsburgh to take somewhere else. Most truckers would rather not ride back empty to Schenectady to get another load from General Electric. The inherent imbalance in flows causes those kinds of problems.

Now, there is, of course, a difference between passenger and freight imbalances. Passenger flows over some period of time will eventually balance out. That is, in our building, everybody goes up in the morning, and everyone comes down at the end of the day. On an annual scale, if we are talking about people who fly from the cold Northeast to Florida, there is certainly an imbalance of flow. In November and December, people tend to fly to Florida. In April and May, they tend to fly back to New England. Over a period of a year, to a first approximation, everybody ends up where they started out, but over the shorter run, say the winter season or the spring season, there are imbalances. Over time, passenger flows do balance out.

However, freight flows do not balance out, because freight takes on different forms at different points in the process. If corn is being shipped from Kansas to New England, the corn doesn't return to Kansas in six months because it gets eaten. So on a freight system, we do not have the averaging out over a long period of time. In the U.S. rail industry, about 42% of freight car-miles are empty, reflecting imbalance in flow and the need to reposition empties. So, another Key Point:

Key Point 24:   Imbalance in flow

24. *Geographical and temporal imbalances of flow are characteristic in transportation systems.*

CHAPTER 9

Transportation Systems: Key Points 25–30

Key Points Continued

Our next Key Point is network behavior. Because the elevator example is such a simple network, it is not very illustrative of this concept. In an elevator, we go up and down. While there are redundant paths in this building network as we noted earlier, there is not the kind of redundancy that one would see in the air network or the highway network or other more complex networks. Understanding and predicting network behavior, however, is at the heart of what we do in transportation systems analysis. Predicting what will happen when we change a network is complex.

You recall that we are talking about this building with three elevators, one servicing floors 1 to 20, the next servicing floors 20 to 40, and the third elevator servicing floors 40 to 60—a simple system. Suppose, that we had a substantial new tenant move in on the fifteenth floor, generating major loads on the elevator servicing floors 1 to 20. What kind of network behavior might we see? What decisions could people make about the way they use the network that would be different from the way they used it before the new tenant moved in on the fifteenth floor? Can anyone think of some ways that individual users of the network might alter their behavior? Yes?

> If you are going to 18, you might go up to 20 nonstop, and then just walk down the stairs two flights.

Yes, this is one option. We might see some adjustments in the use of the network. If these adjustments were substantial, they might conceivably affect other parts of the network.

Although this is a pretty simple example, I would argue that we might have some trouble making predictions about how changes in the demand pattern on this elevator system might change network behavior in subtle ways, because individual people make judgments that they see as optimal for them. In complex transportation networks there are often choke points. The Central Artery is just such a choke point in the Boston area highway transportation system. When an accident happens there, the network behavior that is exhibited can have impacts far removed from the point of the accident.

I can think of a number of occasions where I have called home and spoken to my wife before I leave, and she has said, "You know, I heard on the radio there is a big accident on so and so," and I say, "Why are you bothering to tell me that? It is nowhere near where I go." And it often turns out that I have a slower ride home, even though the accident apparently has no real connection with the route I am taking. There are spillovers and rerouting that takes place. Given the sensitivity of level-of-service to even small changes in volume when the route is close to capacity during rush hour, those kinds of reroutings can have an impact on drivers that are going home via a route that is not close to the accident.

So, network behavior is important. We are concerned about predicting how flows will change when the network changes. When we talk about the concept of capacity in transportation systems, we are talking about the capacity of the links and the nodes in the system. In addition, we have the more complex concept of network capacity. Developing an understanding of how the network behaves as a system, how it operates when it is close to capacity and how changes might be made to make it operate more effectively, is a subtle and complex matter.

Key Point 25: **Network behavior and capacity**

25. *Network behavior and network capacity, derived from link and node capacities and readjustment of flows on redundant paths, are important elements in transportation systems analysis.*

Another network characteristic is circuity—the ratio of an "as-the-crow-flies" distance to distance on the network. In railroad networks, this can be an important factor—they are relatively sparse and distances on the network may be substantially longer than straight-line distances.

The next Key Point deals with the concept of stochasticity in transportation system performance. Here we refer to random effects in the way transportation systems operate and respond to external stimuli. For example, vehicles run out of gas on the Central Artery at rush hour. There may be an infrastructure failure, like a huge pothole that causes the capacity of a particular link to deteriorate because everybody is avoiding the pothole. Weather, such as heavy snow or fog, is another example.

People's behavior is unpredictable. We may have very different utility functions and we may make very different mode choices. Fuel prices and the economy are unpredictable. All these things affect the flows on the networks, as do accidents and the fact that bus drivers may not show up.

Key Point 26:   **Stochasticity**

26. Stochasticity—in supply and demand—is characteristic of transportation systems.

In highway systems, we make a distinction between recurring and nonrecurring congestion. By recurring congestion, we are talking about rush hour congestion. The traffic gets heavy every day at 8:00 a.m. at Fresh Pond, because the number of people that want to use it exceeds capacity.

We also have the concept of nonrecurring congestion, which comes from the random nature of the transportation system such as accidents on the roadway. This kind of congestion is, by definition, very difficult, if not impossible, to predict. Nonrecurring congestion may have a greater impact on delays than the recurring rush hour congestion we see every day.

This has positive implications for the usefulness of technologies like intelligent transportation systems (ITS), which provide real-time information about the system. If we have the sensors in place to locate accidents and the ability to remove them very quickly, we can reduce nonrecurring congestion.

How about congestion on transit and other nonhighway modes?

This varies quite a bit. Transit in this country tends to be relatively uncongested, with the exception of a small number of hours a day. It exhibits sharp peaking in most cities, which leads to congestion for several hours per day.

If we consider the air system, congestion levels depend upon the components that we are examining. There are a variety of ways in which the air system can be congested. It can be congested in the sense that it is hard to get a reservation when you want to fly. There are not enough seats going, say, between here and Chicago between 5:00 and 7:00 in the evening to accommodate everyone. Then there is also airport congestion, the ability of the runways to deal with the number of aircraft in a given period of time during peaks.

One more point on stochasticity and the congestion that it causes: there is a distinction between "peaking" and "stochasticity." Peaking, like daily peaking at the morning and evening rush hours, is systematic. It happens every weekday, and every Friday afternoon going to Cape Cod during the summer. We know there is going to be a peak. Stochasticity refers to the fact that there may be some variation around the average value estimates, which are distinct from peaking. We can conceptualize this by looking at Figure 9.1.

This diagram shows us that on the average at 8:00 a.m., the expected traffic is 3,000 cars per hour. However, because of stochasticity, it is not always 3,000 cars. Sometimes it is 2,500. Sometimes it is 3,200. So, the peak traffic on a particular day is a random variable, but peaking itself is systematic.

The next Key Point:

FIGURE 9.1
Stochasticity.

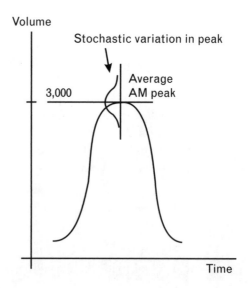

Key Point 27: Transportation, economic development, and land-use

27. *The relationship among transportation, economic development, and*
    location of activities—the transportation/land-use connection—is
    fundamental.

Activities can be industrial or personal, related to a choice of where
to live. Looking at it from a freight perspective, the fundamental
economic theory is that cheap transportation allows a concentration
of production to take place in areas where production can be most
efficiently performed.

Suppose we have two cities A and B. City A can produce some
particular goods very cheaply—let's suppose that it has very cheap
electric power because of the proximity of Niagara Falls—and city
A can produce manufactured goods at low cost because of that advantage. Suppose city B can grow agricultural products very efficiently
because it has very fertile soil, gets a lot of rainfall, and is very sunny (see
Figure 9.2).

If transportation is very expensive between these two cities, the
cities will produce, say, printing presses and corn for their own consumption. If we had efficient and inexpensive transportation services
between these two cities, what would happen? Now the cost of transporting printing presses from A to B is small. We can produce them
cheaply at A and it is expensive to produce them at B because, suppose,
we do not have a cheap source of power. Now, because the transportation cost for shipping printing presses from A to B is small, we can
produce printing presses at A and ship them to B less expensively than
they can be produced at B in the first place.

Conversely, the corn at B can be shipped at low cost to A, where
corn does not grow well. So we can see a concentration of economic
activity at A in manufacturing with economies of scale and at B in
agriculture, also with economies of scale, enabled by the development

FIGURE 9.2 *Two cities*
and their economies.

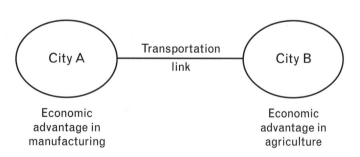

of transportation services between those two cities. It could, in principle, produce economic benefits at both A and B. Further, with higher transportation volume between A and B, we obtain transportation economies of scale as well.

Shifts in Activity Base Related to Transportation

Manheim [1] talks about interactions between transportation network flows and the activity base. He discusses the importance of transportation investment and the shift of activities to areas of cheap transportation. Certainly the article of faith is that transportation investment, if intelligently made, is something that can have a positive impact on economic activity and on quality of life. We have viewed this as a tenet of transportation for some years. It does not always work. If we make a poor investment in infrastructure, we may be worse off because those same resources could have been used in some better way. Nevertheless, the idea that providing transportation infrastructure can be a positive economic force is important.

Yes?

How about the journey to work and urban location patterns?

Clearly, the development of transportation infrastructure and the resulting personal accessibility affects choices about where to live. Urban planners talk about the "rule of 45 minutes" as something they argue has been relatively constant over human history. If one looks at maps of ancient cities in which the fundamental mode of getting through the city was walking, people tended to settle in a pattern such that it was about a 45-minute trek to get from where you live to wherever it was you were going to earn your daily bread. Ancient cities tended to take about 45 minutes to 1 hour to cross on foot.

As transportation technologies improved, the size of cities and metropolitan areas grew and the "rule of 45 minutes" has held relatively constant. When we went to horse and carriage, when we went to automobiles, when we went to higher speed trains, the size of the city grew because transportation enabled people to live further away from their worksites.

In Japan, for example, there are Shinkansen trains operating at 150–175 mph. There are people who routinely commute to downtown Tokyo on the Shinkansen that live 100 miles from Tokyo; it is a routine, although expensive, commute. Those commuters are there as fast as someone living in Greenwich, Connecticut working on Wall Street can get to work—a distance of 30 miles. Now, these long-distance Shinkansen users pay a rather substantial price for that transportation; Shinkansen travel is expensive on a per-mile basis. But the fact is that the basin from which Tokyo can attract users has expanded substantially, at least along the Shinkansen corridors. On the scale of New England, this would be equivalent to living in Portland, Maine and working in Boston, which would be very difficult because high-speed transportation services are not available.

Measuring Transportation System Performance

The next Key Point deals with performance measures for transportation systems. The fundamental question is how we go about measuring the quality of performance on a transportation system. This may seem, on the face of it, to be very simple. In practice, it is nontrivial.

What Do Your Customers Perceive?

There are several different concepts we want to capture in a performance measure. The first concept is that we want to relate those performance measures to the level-of-service of the system as perceived by our customers. Realize that these performance measures are multi-dimensional. For example, we do not just make travel judgments based only on travel time, but on other parameters as well. Let me emphasize that our ways of measuring performance have to relate to the ways in which our customers make decisions about whether to use us or our competitors.

Performance Measures and Cost

A second concept is that performance measures have to, in some way, relate to costs of operations and revenues derived from operations of those systems. This applies whether we are talking about a profit or a nonprofit organization—whether we are talking about the Union Pacific Railroad in the business to make money, or the MBTA, which has a substantial public subsidy. There are, in either case, financial measures that we use—capturing revenues and costs—that measure the efficiency with which we are producing transportation services and the attractiveness that those transportation services have in the marketplace. Cost is a reflection of how efficient we are at producing transportation service; revenue is a measure of how good we are at attracting customers.

System Versus Component Performance

A third concept: recognize the difference between overall system performance as contrasted with component performance. Typically, the customer sees overall performance; if we are using the Union Pacific Railroad to ship our goods, we do not observe and we probably do not care very much about what happens between point A and point B on the complicated Union Pacific rail network. What we, the customer, care about is travel time, service reliability—these are measures that affect us directly. The numbers of times our shipment was handled en route may, in fact, affect that performance, but we do not directly care about it. All we care about is the performance as we see it.

While the customer sees overall performance, the operation is often managed on a component basis. It is very difficult to manage the system in its entirety. Rather, we manage components of the system, such as freight terminals, vehicle fleets, and so forth. The hope is, if we do a good job of operating the components, the system as a whole will operate well. Usually, it is a necessary condition that component operation be effective for system operation to be effective, but it is very often not a sufficient condition. We can have a poorly meshed system in which the components are operating well, but the system performance is still poor.

Performance Measures and Behavior

Realize that the way in which we measure the performance of those components is going to have an effect on how the person operating that component makes decisions. For example, consider a bus driver on the Number 1 bus on Massachusetts Avenue. Suppose we measure that bus driver on schedule adherence. "We want you to leave on time and we want you to get to the other end of the line on time. We are going to penalize you if you do not, and we are going to give you credit if you do; schedule adherence is the way that we are measuring you." That bus driver would operate in a particular way.

Now suppose that we decided that we were going to measure that bus driver on the revenue collected during his or her run. I would suggest that bus drivers would probably operate differently under those two circumstances. If schedule adherence is what he is being measured on, he will do what is necessary to get to the other end of the line on time, perhaps even if it means skipping the last four stops. On the other hand, if revenue maximization is the performance measure of interest, he may get the bus as full as possible and perhaps stop for people between stops if he has to, just to load up the bus.

Now, obviously, the goal is to achieve some balance between the kinds of measures we use, but the point is illustrated. The behavior of people who are managing and operating the transportation system is affected by the performance measures by which they are being measured. If we are talking about the manager of a railroad terminal, we could imagine all sorts of ways of measuring his or her performance that would push him or her to operate differently. Suppose we tell this terminal manager, "We are a long-train railroad. We believe the economical way to run railroads is to run long trains. We are going to measure you on how long your trains are. If you run short trains, you are in big trouble."

So, that manager is probably going to delay trains for incoming traffic so that he or she can run long trains. If, on the other hand, we tell the terminal manager, "You are going to be measured on the cars that are left in your yard at the end of your shift—the more there are, the worse your performance was," he or she is going to move a lot of cars out of that yard even if he or she has to do it on very short trains.

If we tell her it is schedule adherence, she is going to operate one way; if we tell her it is yard resources used, like switch engines and crews, she is going to operate in quite another way. We are going to see differences in operating strategies that are based upon the kinds of incentives that we offer people.

People (and organizations) respond to incentives.

Network Performance

The question of performance measures in transportation systems is made more complicated by the fact that transportation operators have to operate within the context of an overall network over which they may have very modest control or, no control. Let us suppose we have a particular incentive system for our terminal manager. Now, suppose that the trains coming into our terminal had arrived several hours late because they were dispatched late by other terminal managers over whom our terminal manager had no control. How do we fairly measure the performance of our manager? It is not an easy question. People have been working on this question for a long time. How do we measure the performance of a manager who is forced to respond to issues that are beyond his or her control?

Key Point 28:   Performance measures
28. Performance measures shape transportation operations and investment.

One of the important tensions we have in controlling networks and developing performance measures is between people who feel centralized operations can be performed optimally and people who feel that the on-the-ground, real-time knowledge of what is going on is critical as well. If the centralized control really knows all the information about the system in real-time then, in principle, overall optimization is feasible. But, in practice, it is impossible for the central control to have all that information. This leads us to our next Key Point.

Key Point 29:   Balancing centralized with decentralized decisions
29. Balancing centralized control with decisions made by managers of system components (e.g., terminals) is an important operating challenge.

Our final Key Point goes back to the components we introduced earlier in this discussion. Those components, including infrastructure, vehicles, and control systems, must operate effectively as an overall system. Judgments made in design and operation of these components must take their interactions into account.

In the elevator example, the elevators must be sized appropriately for the shafts. Other examples are the need to dredge ship channels and harbors to accommodate larger ships; assuring that trains can safely pass under highway overpasses; and, designing highways such that they can carry heavier contemporary trucks.

Key Point 30:   Integrality of vehicle/infrastructure/control systems decisions
30. The integrality of vehicle/infrastructure/control systems investment, design, and operating decisions is basic to transportation systems design.

The following is a summary of our 30 Key Points.

....

Summary of Key Points

1. People and organizations alter behavior based on transportation service expectations.

2. Transportation service is part of a broader system—economic, social, and political in nature.

3. Competition (or its absence) for customers by operators is a critical determinant of the availability of quality transportation service.

4. Analyzing the flow of vehicles on transportation networks, and defining and measuring their cycle, is a basic element of transportation systems analysis.

5. Queuing for service and for customers and storage for vehicles/freight/travelers are fundamental elements of transportation systems.

6. Intermodal and intramodal transfers are key determinants of service quality and cost.

7. Operating policy affects level-of-service.

8. Capacity is a complex system characteristic affected by: infrastructure, vehicles, technology, labor, institutional factors, operating policy, external factors (e.g., clean air, safety, regulation).

9. Level-of-service = f(volume); transportation supply. As volume approaches capacity, level-of-service deteriorates dramatically— the "hockey stick" phenomenon.

10. The availability of information (or the lack thereof) drives system operations and investment and customer choices.

11. The shape of transportation infrastructure impacts the fabric of geo-economic structures.

12. The cost of providing a specific service, the price charged for that service, and the level-of-service provided may not be consistent.

13. The computation of cost for providing specific services is complex and often ambiguous.

14. Cost/level-of-service trade-offs are a fundamental tension for the transportation provider and for the transportation customer, as well as between them.

15. Consolidation of like-demands is often used as a cost-minimizing strategy.

16. Investments in capacity are often lumpy (e.g., infrastructure).

17. The linkages between capacity, cost, and level-of-service—the lumpiness of investment juxtaposed with the hockey stick level-of-service function as volume approaches capacity—is the central challenge of transportation systems design.

18. Temporal peaking in demand: a fundamental issue is design capacity—how often do we not satisfy demand?

19. Volume = f(level-of-service); transportation demand.

20. Level-of-service is usually multidimensional. For analysis purposes, we often need to reduce it to a single dimension, which we call utility.

21. Different transportation system components and relevant external systems operate and change at different time scales (e.g., short run—operating policy; medium run—auto ownership; long run—infrastructure, land-use).

22. Equilibration of transportation supply and demand for transportation service to predict volume is a fundamental network analysis methodology.

23. Pricing of transportation services to entice different behavior is a mechanism for lowering the negative externalities caused by transportation users on other transportation users and society-at-large.

24. Geographical and temporal imbalances of flow are characteristic in transportation systems.

25. Network behavior and network capacity, derived from link and node capacities and readjustment of flows on redundant paths, are important elements in transportation systems analysis.

26. Stochasticity—in supply and demand—is characteristic of transportation systems.

27. The relationship among transportation, economic development, and location of activities—the transportation/land-use connection—is fundamental.

28. Performance measures shape transportation operations and investment.

29. Balancing centralized control with decisions made by managers of system components (e.g., terminals) is an important operating challenge.

30. The integrality of vehicle/infrastructure/control systems investment, design, and operating decisions is basic to transportation systems design.

Having completed our Key Points discussion as an introduction to transportation systems, we now proceed to discuss models and frameworks that are useful in the transportation context.

REFERENCE

1. Manheim, M. L., *Fundamentals of Transportation Systems Analysis*, Vol. I, Cambridge, MA: The MIT Press, 1979.

CHAPTER 10

Models and Frameworks

....

Models and Frameworks: An Introduction

Next, we present the idea of models and frameworks. We do so in the context of transportation; however, this philosophy will be useful in other fields as well. We start by distinguishing between models and frameworks. For our purposes, "models" are mathematical representations of a system. By "frameworks," we mean a qualitative organizing principle for analyzing a system. We can use both models and frameworks to do analysis; in the case of a model, the results will be quantitative (e.g., numerical or in equation form); in the case of a framework, the results will be in qualitative form (e.g., typically in the form of words).

I emphasize that both of the above are analyses, although stylistically different. As we proceed, we will note a number of examples of both models and frameworks. We start by asking the following question:

What is our function as transportation professionals?

What is it that we, as transportation professionals, really have to do, now that we understand some of the fundamental underlying premises of transportation systems and how they work? There are a number of functions we could talk about. We could talk about designing better transportation systems. We could talk about using resources, financial and otherwise, effectively in a transportation context. We could talk about operating transportation systems optimally. Finally, we could talk about maintaining transportation systems efficiently.

These are all laudable goals, but realize that there are some value-laden words in the descriptions of what we as transportation people do: "better, effectively, optimally, efficiently," may very well depend on your point of view. What optimization of a system really means depends substantially on whether you are the CEO of the Union Pacific Railroad, or the chairman of the Surface Transportation Board, or the director of the Environmental Protection Agency. In any case, regardless of your point of view, you need to develop a structure within which you can look at ways of doing things better, effectively, optimally, and efficiently.

A Structure for Transportation Systems Analysis

Figure 10.1 shows a flowchart, which not only applies to transportation, but to systems analysis in general. The idea, starting from the top of the flowchart, is that we begin by searching for alternatives for the real world system. We think about better ways of operating the bus system, of investing in the airline, or of building the highway system. After

FIGURE 10.1   *A systems analysis framework: a first look.*

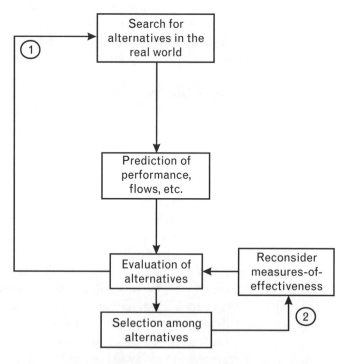

Loop 1: Does the evaluation suggest other alternatives?
Loop 2: Are the measures-of-effectiveness appropriate?

coming up with alternatives, we try to predict the performance of the system for that particular alternative (as shown in the next box down). We then develop some measures of effectiveness—that is, what are some figures of merit for us to use in deciding whether our system is operating effectively? We look at the predictions for performance for various alternatives, go through an evaluation of those alternatives, and then we select one or more alternatives that make sense. This is the fundamental structure within which we operate.

The flowchart has several loops. The feedback loop labeled 1 goes back from "evaluation of alternatives" up to "search for alternatives." After evaluating some alternatives, we may discover—as a result of that evaluation—other alternatives that we may want to consider. We again go through some kind of iteration, and our evaluation of an alternative—that is, our interpretation of the result of modeling—may lead us to some alternatives we had not thought of before.

Loop 2 goes from "selection" back to "measures of effectiveness." It is plausible that one may decide, based upon the selected alternative, that perhaps the initial measures of effectiveness were not particularly good.

The Subtlety in Choosing Measures of Effectiveness

Here is an example to illustrate this point. Consider the stylized representation of an urban area in Figure 10.2.

Suppose we have all the high-income people in the Northwest and we have all the low-income people in the Northeast. We are trying to decide whether we should expand the highway system to provide for the high-income people, as opposed to replacing a very modest bus system from the low-income area with a more effective rail transit system.

FIGURE 10.2
An urban area.

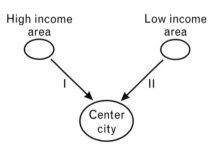

One classic measure of effectiveness that we use in analyzing transportation systems is time savings; value-of-time is in that equation. How much time—and therefore money—is saved as a result of the deployment of a particular kind of transportation system? Traditionally, time is valued differentially, often as a function of income. In transportation planning, we usually use some fraction of hourly wages—typically something like 70% or 80%, to measure the value of time. Now obviously, people earning $500,000 per year have a much higher hourly wage than lower income people.

So, when we go through the analysis based on value-of-time, we might come to the conclusion that, "Well, the obvious thing to do is to build the highway system and save time for the high-income people rather than for the low-income people." After further evaluation, we might decide that perhaps the measure of effectiveness that we were using was not particularly appropriate. So there is a feedback loop between what gets selected and the development of measures of effectiveness.

Abstraction of Real World into Models or Frameworks

Earlier in Figure 10.1, we left a gap in the flowchart between "search for alternatives" and "prediction of performance." This gap is for the abstraction of the real world into a model or a framework (see Figure 10.3).

Abstracting from the real world into a model or a framework is a very important step because very often—virtually all the time—we are unable to do experiments on the real world. It is unlikely that the Union Pacific Railroad or Delta Airlines is going to say to some researcher, "So, we've heard you have this great new idea about how to run the system. Why don't you just take it over for a week and run some experiments?" This is very unlikely.

Models

Let's focus on models and the quantitative approach first. Later, we will introduce frameworks and the qualitative approach.

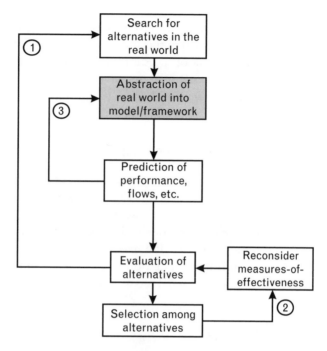

Figure 10.3
A systems analysis framework: a second look.

Loop 1: Does the evaluation suggest other alternatives?
Loop 2: Are the measures-of-effectiveness appropriate?
Loop 3: Is the abstraction good at predicting?

When we cannot experiment in the real world, we must build a mathematical abstraction of reality on which we can do our experiments. We build a model of the Union Pacific and perform some experiments. We try various alternatives and come up with ways to improve the operation of the Union Pacific Railroad. We do experiments on the abstraction—the model—not on the railroad itself.

Is the Abstraction Any Good?

So, this leads to another loop. Loop 3 asks the question of whether the abstraction that we have constructed is useful in making predictions. That is, we need an abstraction that represents the real world such that when we change a variable, the outputs or predictions from the abstraction—the model—change in a reasonable way. If the predictions do not make sense, we need to go back and change our model so that it conforms to our view of reality.

We have these three loops: thinking about new alternatives after we have evaluated the alternatives we came up with at first; thinking about new kinds of measures of effectiveness after doing the initial selection; and looking at whether the model is, in fact, helpful in predicting performance of the real-world system.

Why Are We Modeling?

Even at this early stage, a reasonable question to ask is "Why are we modeling?" What kinds of questions do we have in mind when we model and how might those questions suggest how we choose to develop our model? My model of the Union Pacific and your model of the Union Pacific may be very different. We have to go back to the fundamental question of why we are modeling in order to know what kind of a model we want to build.

Insight

There are various reasons to do modeling. First, one reason is the development of models for insight: learning more about system performance by looking at model representations of the real world. I was reading a book recently on *Science and Ideas* and there was an article by a physicist named Perry Bridgeman, who was talking about physics, not transportation, of course. He made a comment close to some of my own thinking on modeling. He said,

> *We may gain insight into complex situations by first understanding simpler situations resembling them.*

We do this all the time. Representing the world in all its detail is very hard, very confusing, and very confounding. So, a technique that many people use is to come up with a model that they know is wrong (all models are wrong—however, some are useful) but is very simple. Of course, models do not account for all of reality; however they provide us with some insight into the way systems perform. We would not actually use a simple model to make specific decisions about how to run the Union Pacific Railroad; to do that, we need to take into account more real-world complexities. But, we may gain some good insights into how to run the Union Pacific Railroad by looking at a much simpler (wrong) model. For obvious reasons, we call this an "insight model."

I drew a branch on our flowchart to an insight box. Before I begin considering what to do in the real world, I use simple models that allow me to gain some insights into the way in which complex systems operate (see Figure 10.4). In many instances, when I start a new problem, that is the first approach.

Choosing the Best Alternative—Optimization

A second reason that we model is straightforward: it is simply to choose the best alternative for operating a system, for investing in a system, for maintaining a system, and so forth. Sometimes the model allows us to choose the best alternative directly. The model itself produces an optimum. We build a model, input the data and we solve it; the number that results is optimal, by the very structure of the model.

FIGURE 10.4
A systems analysis framework: a third look.

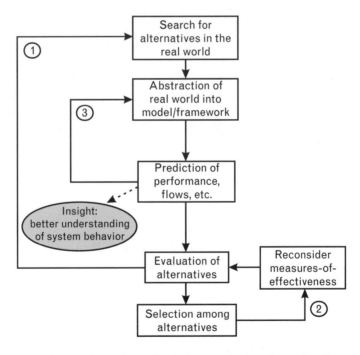

Loop 1: Does the evaluation suggest other alternatives?
Loop 2: Are the measures-of-effectiveness appropriate?
Loop 3: Is the abstraction good at predicting?

Getting an optimum directly is not always possible. Often, rather than generating an optimum, we use the model as an aid to human decision making. The model gives us, say, three good ways of operating a transportation system; since we cannot describe our measures of effectiveness precisely, we take these three alternatives and use whatever is implicit in our head (but not in the model) about what is good and what is bad about these alternatives to pick one. We use common sense, our sense of the pragmatic, and perhaps political realities.

System Operation

A third reason for modeling is to actually operate the system. We want to use the model in real-time to operate the Boston traffic light synchronization system. We want to actually produce results that drive the system. Now, this kind of environment presents some real constraints on the kinds of models we can have. If we want to be able to operate the actual system in real-time, based upon the outputs of a model, the model has to go fast enough to work in real-time.

The necessity for real-time modeling is a chronic problem in weather prediction. Meteorologists have complex, highly detailed models that purport to be able to predict the weather in Boston an hour from now; maybe they can and maybe they cannot. The real problem, however, is that it takes 24 hours to produce the predictions. So, to use those results in real-time is obviously not possible. The same thing goes on in transportation. If we are going to have a model that tells us how to run the traffic system—how to change the traffic light settings in five minutes—it better not take me 25 minutes to get the answer out of that model. If so, the model is of little value for real-time applications.

Learn from Model Building Process

A fourth reason for modeling is to learn about the real-world system from the actual process of building the model—the actual process of doing the abstraction. I have found this time and time again to be a very effective aid to system understanding. It is analogous to writing a computer program. When we are writing a procedure for a machine that takes everything literally, we have to think through, in great detail, a step-by-step algorithm for how the procedure is to be performed. When we are abstracting reality we have to think quite clearly about what to include, what not to include, how the system operates, what will happen if we make investments in particular ways, and so forth. So learning more about the system simply by building the model is important. In some cases, it may turn out that building the model is sufficient. The knowledge gained in constructing the model may provide us with enough insight to make the necessary decisions about the real-world system.

Modeling for Negotiation

A fifth reason is what I call modeling for negotiation. Very often in the world of transportation, and more generally in systems analysis, we as transportation professionals operate in competitive situations. It may be a competition between various transportation companies, two railroads, a railroad and a trucking company, or competition between management and labor. It may be a competition between the vice-president of operations, who wants to minimize cost, and the vice-president of marketing, who wants to maximize service. When we have these kinds of competition, we must have some basis on which to negotiate. Once a basis for negotiation is established, we can sit down and say, "Here's what I want and here's what you want. I'll give you this and you give me that. Let's work towards something that actually makes sense for both of us in the real world."

We have discovered that models can provide a framework for negotiating about how a system should be deployed. This is a very useful modeling application. A model called the service planning model (SPM) was developed by Carl Martland and J. Reilly McCarren in the late 1970s [1]. It is still being used by a number of railroads [2]. It is a simple representation of a railroad network—much simpler than the highly-detailed, micro-simulations of railroad networks that are available. It is, however, close enough to reality to model and predict the basic trade-offs between cost and level-of-service on the rail network being studied.

The SPM has been used by railroad managers to facilitate an effective negotiation between operating people and marketing people. They sit down and say, "Let's actually see what happens if we add this train. What happens to costs? What happens to service? Let's see what happens if we decide to run twice daily service with train lengths of 50 rather than once-daily service with train lengths of 100. Here is what happens to cost; here is what happens to service." And the

VP-Marketing says, "I can live with this if you can live with that," and back and forth they go. The optimization takes place through a negotiation process between people. The model is basically the mechanism with which they negotiate.

In the urban setting, a good example of this "modeling for negotiation" is the research of Prodyut Dutt [3]. He focused on how planners work in developing countries—cities like Bangkok and Bombay—versus the way urban planners work in the developed world—cities like London and Chicago. The essence of his research looked at the differences in planning methodologies necessary in the two environments.

In part, Dutt developed "models for negotiation." He talked about the idea of transparent models, where rather than shielding the outside world from what was going on within that "black box" model, planners could look into the model, in a sense, and clearly see the assumptions driving the trade-off about location of corridors in urban areas, for example. It was a very simple and transparent approach to urban modeling compared to what planners see here in the United States. Although simple, Dutt's model was a very effective method for negotiating land-use and transportation decisions because the model lets planners see the macro-results of various assumptions and selection criteria. So, the model can be an effective mechanism for negotiation.

So, providing simple models to allow negotiation is very helpful and important. Of course, in any circumstance where we have a simple model, we must be sure that it is not so simple an abstraction of the real world that, while malleable and understandable, it is not particularly useful in understanding what is going on. I would suggest that this is more art than science. It is not an easy thing to determine when the benefits of a simple model—modest data and execution requirements—are outweighed by the lack of usefulness of the prediction.

The level of precision in the results a model can produce is important as well—depending on the application. If we plan to use the model for an application where precise results are central to success—such as computing the re-entry path for a manned spacecraft—we need to build a model with a level of detail and world view appropriate to the task. On the other hand, if all we need is a first-order feel for a situation—such as whether interest rates will rise (or fall), but we are not concerned with how much—a less detailed approach may be appropriate. The general statement is:

The modeling approach depends on how we are going to use the results.

The Model Is a Shaper of Your World View

How we abstract the real world into a model has a way of controlling the way we think about the system we are studying. Once we have decided that we are going to use a simulation model, or once we have decided we are going use a network equilibrium model, we have made a decision about our mindset on that system. So, it is very important that when we come up with a model, we are coming up with one that is consistent with the way we want to think about the system.

Modeling ideas shapes the way we look at the system, rightly or wrongly. The old joke is that when all we have is a hammer, everything looks like a nail. If what we are interested in is system dynamics and we love system dynamics models, we think of the world in terms of system dynamics, when perhaps that world view is not helpful. The same is true of a variety of other modeling types. Falling in love with a particular modeling approach, and then using it for everything—sometimes inappropriately—is a common error in modeling.

Thinking about abstractions and thinking about how we are going to represent the real world is a very complex and subtle process. On a lighter note, there is a joke that has been around for years about two buddies, one engineer and one physicist, and they were taking a vacation together on a dude ranch. They were talking about the relevance of engineering versus the relevance of physics. Each of them was arguing that their own discipline was the way to represent ultimate truth and could answer questions that people wanted to have answered.

So the engineer says to the physicist, "Well, let's see how good you really are at this. Do you see that cow sitting out there in the field chewing grass? I'd like to see you use your physics principles to come up with an energy balance model for that cow. The sun impinges on the grass, which the cow is chewing; there is waste product. I'd like to see you come up with a model from your physics discipline that gives you a representation that would be helpful." The physicist says, "I can do that. Just give me an hour or two and I'll come back with a very useful model." They have lunch and afterwards the engineer says, "Well, I'd like to hear about your model." The physicist says, "Well, I've got it all set for you. Let's start by assuming a spherical cow."

I suspect that it is not a joke at all. Spheres are a representation of the physical world that happen to be very amenable to quantitative analysis. The point is that representing the cow as a sphere may be exactly the kind of representation for the kind of question that was being asked. That is where the judgment—the art—comes into the picture.

Another example of the way in which our models strongly affect the way that people think about systems—and may constrain us from coming up with the "right" answer—comes from economics. There was a series of articles by David Warsh in *The Boston Globe* that discussed the history of economic thought over the last 40–50 years. He talked in one of his articles about the relationship between mathematical representations and reality. Here is a quote from one of his articles. We will not discuss the details, but he is talking about imperfect competition as a way of thinking about how economic systems operate. He says, "Consider the new emphasis on various forms of imperfect competition that is now sweeping economic theory. This was a phenomenon which for many years had been all but banished from economic texts as being uninteresting simply because it was mathematically intractable."

Basically, what he is suggesting is that since the mathematics of imperfect competition at the time (back in an earlier generation of economic thought) was intractable, the idea of imperfect competition was rejected as an explanation of the real world. We have to be very cautious as we adopt a model—making sure that in order to be able to solve the model, we are not making a representation that is in fact wrong, just because we cannot do the math on the more appropriate representation. To illustrate, here is another (lighthearted) story.

This story is about a man under a street lamp on a dark street who is looking around on the ground; a policeman comes along and says, "What's the matter?" and the man says, "Well, I lost my keys." The policeman says, "Well, where did you lose them?" The man points up the street and says, "I lost them up there somewhere." The cop says, "Well, then, why are you looking under the street lamp?" and the man says, "Oh, the light is much better here." In a sense, that story represents some modeling approaches. The light is much better here. We know how to do simulation; we know how to do system dynamics. So let's do it; perhaps it has little to do with reality, but "the light is better here."

Modeling Approaches

Some modeling approaches that have proven useful over the years in transportation systems analysis include:

- Network analysis;

- Linear programming;

- Nonlinear programming;

- Simulation;

- Deterministic queuing;

- Probabilistic queuing;

- Regression;

- Neural networks;

- Genetic algorithms;

- Cost/benefit analysis;

- Life-cycle costing;

- System dynamics;

- Control theory;

- Difference equations;

- Differential equations;

- Probabilistic risk assessment;

- Supply/demand/equilibrium;

- Game theory;

- Statistical decision theory;

- Markov models.

There is no shortage of modeling approaches to use to study transportation problems. Some of them may be quite germane, and again the art is knowing what to use and when.

Getting Answers From Models

Coming up with a mathematical representation in any of these modeling approaches and making answers come out of the model are very different things. We can develop, for example, very complicated probabilistic queuing networks to study transportation problems. We can formulate the equations and write them down. The equations can be elegant and beautiful, but getting an answer—taking the probabilistic representation that we have come up with and saying, "I am going to take this set of equations and from it is going to come an answer," is quite another question. So some of the art of what we are doing is choosing the models we use from the viewpoints of their ability to represent reality and their ability to produce answers.

It is not all that hard to produce elegant mathematical models of transportation situations. The hard part is producing elegant models that we know how to solve—where we can stare fixedly at the equations or use a computer package and solve them and produce an answer. If we have a set of linear differential equations, we know we can solve them. But if we have a complex, highly interconnected network with probabilistic behavior, formulating a model and obtaining an answer are two very different things. So, knowing the state of the art in modeling is very important—what models can be solved?

Coming up with a modeling approach that cannot give us a result—an answer—is not helpful. We can relate this to our earlier modeling discussion. Between the steps of abstracting the real world into a model/framework and prediction of performance, we can insert a new step—can the model be solved; can it produce results? If so, we are fine, assuming the predictions are adequate for the purpose. If not, we need to go back and re-abstract into a model we can solve, but that is still a good representation of reality; this we show as Loop 4 (see Figure 10.5). Usually, we do this all in one step, since experienced modelers know what models can or cannot be solved.

Models Versus Frameworks

Earlier, we mentioned models and frameworks. We need to re-emphasize the distinction between models and frameworks. This is adapted from some of the thinking of Professor Michael Porter of the Harvard Business School.

Frameworks

When we talk about a model, we are talking about some kind of mathematical representation of reality that is quantitative in nature. When we talk about a framework, we are generally talking about a qualitative

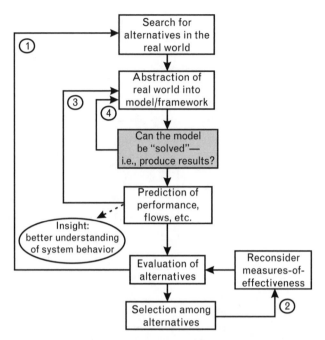

FIGURE 10.5
A systems analysis framework: a fourth look.

Loop 1: Does the evaluation suggest other alternatives?
Loop 2: Are the measures-of-effectiveness appropriate?
Loop 3: Is the abstraction good at predicting?
Loop 4: Develop new abstraction

view of a complex system. It is a way of thinking—a way of organizing our thinking about a complex system—not necessarily numerically, but in an organized form. A framework may be a set of concepts or organizing principles useful in understanding a complex system. It is descriptive and qualitative rather than mathematical and quantitative.

We have seen several of these frameworks in relatively simple forms so far. For example, we talked about a framework for thinking about what we were going to present in this book. We described a three-dimensional matrix that is, in a sense, a framework for thinking about—for organizing knowledge about—a transportation system (see Figure 10.6).

Porter's Framework—Strategic Comparative Advantage

An excellent example of such a framework is one of Porter's. He has done a lot of work in what he calls strategic comparative advantage

FIGURE 10.6
*Transportation systems
characterization.*

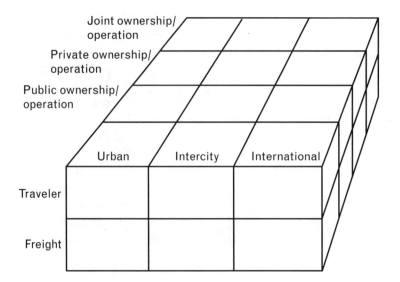

[4]. He is typically talking about comparative advantage, say, between firms, or perhaps between countries—entities that are, in some sense, competing.

In the case of a firm, he comes up with what he calls a framework for analysis. The details of it are not important for us here, but he talks about the way in which one goes about thinking about comparative advantage—he talks about characterizing rivalry among firms, the bargaining powers that exist in that marketplace between the buyers and the sellers of services, the threat of new entrants into the marketplace, and the threat of substitute products. This is simply an organizing principle for looking at a competitive situation.

We could use it to look at railroad transportation or to understand the competitive relationship between Western Europe and Japan or the United States and Southeast Asia. It is simply a way of thinking about a problem.

Another framework is re-engineering. Some of you may have read the books by Hammer and Champy on re-engineering as a way of thinking about reorganizing organizations and processes therein [5]. We could argue organizational re-engineering is a framework for analysis. It is basically a framework for thinking about, understanding, and solving a problem.

When we talk about models, we are talking about mathematical constructions. When we talk about frameworks, we are talking about conceptual understanding. Both models and frameworks are analyses tools. In broad terms, analyses done with models are mathematical—they are equations, they produce numbers. Analyses done with frameworks tend to be qualitative—they produce insights, they produce ideas, they are a way in which we can organize our thinking about complicated kinds of problems. Because this is frequently misunderstood, it should be emphasized that both models and frameworks can be used for analysis.

Simplicity Versus Complexity

Here is a matrix that shows some choices: we can take a system and we can model it mathematically, or we can look at a framework for understanding the system qualitatively (or conceptually); in either case, we can analyze the system at a simple or complex level (see Figure 10.7).

We can have simple mathematical models or complex mathematical models; the same applies for frameworks. The question we have to ask is, what is useful? What is helpful in understanding the situation?

We can have both "useful" and "not useful" in any of the boxes of the matrix. Complexity (or simplicity) does not make something useful whether it is quantitative or qualitative. We have to think very carefully about the level of detail and complexity we are going to use to represent the system, whether we are thinking about it quantitatively or qualitatively. Just because we quantify something does not make it inherently right. Furthermore, just because we cannot quantify something does not mean it cannot be usefully and analytically studied. Frameworks can be organized, systematic, and normative—in short, useful.

Now, clearly, quantitative representations are very useful. Quantitative models can be extremely helpful in understanding, designing, and operating transportation systems. Broadening our discussion to

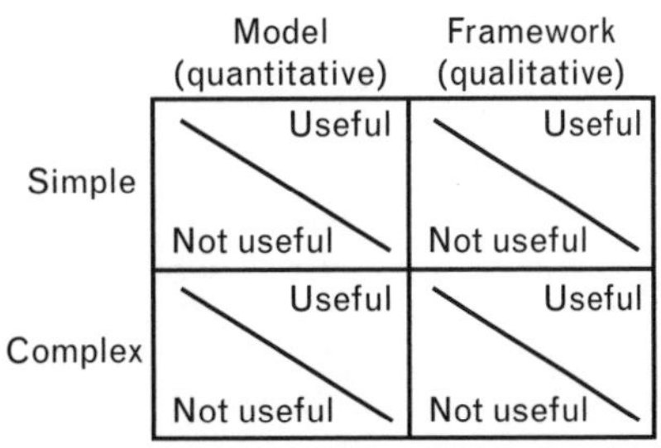

FIGURE 10.7 *Simple versus complex models and frameworks.*

include conceptual frameworks to study transportation issues, we should recognize that there are nonquantitative ways of thinking about transportation that can be more useful in some cases than quantitative methods. Any questions?

If we want to study a transit system, should we use models or frameworks?

I suspect we would want to have both kinds of representations. We would have mathematical models to help us figure out how to do scheduling, but we might need a framework—an organizing principle—to understand how to go about improving the flows of funds from the Commonwealth of Massachusetts to the MBTA, probably a qualitative problem. Studying the flow of funds would require an organized set of principles, a framework for understanding that could give us more insight into how to go about improving that process. So both models and frameworks would apply to transit systems. This is a good example because it illustrates that we can use both models and frameworks in studying a system: highly mathematical approaches for parts of it and more qualitative ways of assessing other parts.

What about predicting demand for transportation services?

Again, the approach includes both models and frameworks. For example, in the work that Moshe Ben-Akiva and Steven Lerman do, highly quantitative mathematical representations of behavior are used.[1] They measure what people do; they ask people what they will do, they develop datasets, and they try to predict what people will do if a new transportation mode is added, or if the quality of service is greatly improved on transit, whatever it may be.

But, at the same time, we have qualitative frameworks for understanding how people make decisions. Professors Ben-Akiva and Lerman also talk about what could be considered a framework for analysis for transportation demand. They look at short, medium, and long-term decision-making as a basic framework for thinking about decisions. We make long-term decisions about where to work. We make short-term decisions about routes (e.g., do we go through the back streets of Cambridge or by way of Route 2 today?). Both models and frameworks

---

1   Ben-Akiva, M. and S. R. Lerman, *Discrete Choice Analysis: Theory and Application to Travel Demand*, Cambridge, MA: The MIT Press, 1985. This is the definitive reference on transportation demand.

are ways of thinking about how transportation demand works. It is both quantitative and qualitative.

Let's go on to discuss modeling concepts in more detail, focusing on models that can be useful in transportation.

REFERENCES

1. McCarren, J. R. and C. D. Martland, "The MIT Service Planning Model," *MIT SROE*, Vol. 31, December 1979.

2. Van Dyke, C. and L. Davis, "Software Tools for Railway Operations/Service Planning: The Service Planning Model Family of Software," *Computer Applications in Railway Planning and Management*, Boston, MA, 1990, pp. 50-62.

3. Dutt, P., "A Standards-Based Methodology for Urban Transportation Planning in Developing Countries," Ph.D. Thesis, Department of Urban Studies and Planning, Massachusetts Institute of Technology, Cambridge, MA, September 1995.

4. Porter, M., *Competitive Strategy: Techniques for Analyzing Industries and Competitors*, The Free Press, 1980.

5. Hammer, M. and J. Champy, *Re-engineering the Corporation: A Manifesto for Business Revolution*, Harper Business, 1993.

CHAPTER 11

Modeling Concepts

....
Introduction to Models

We now discuss modeling concepts to emphasize what is important in building transportation models. If we are thinking, for example, about developing a transportation model of the Boston metropolitan area, we can relatively quickly come up with any number of modeling issues we would have to deal with in representing transportation in Boston.

For example, we have to decide about the scale at which to study Boston. How many nodes do we represent Boston with? 5? 50? 500? 5,000? How many links? At the link level, we have modeling choices: do we represent traffic as a set of individual vehicles (discrete model) or as a flow (continuous model)? At the node level, we can represent a bus terminal, for example, as a deterministic delay function or we can model it as a complex micro-simulation, which gives us a probabilistic representation of delay for each vehicle. What are the implications of these decisions?

How many different modes do we consider in developing this model of Boston transportation? It is clear that we would have separate modes for automobiles and transit, but do we need a separate mode for walkers and bicyclists? Should these modal models all be interconnected? Or, should we have independent models representing each of these modes? Should we consider the flows on the networks as static, or should we represent flows dynamically—as a function of time?

Should we consider short run or long run issues? Do we allow for changes in land-use? Do we consider equilibrium? We talk about equilibrium as a way to think about transportation systems, and we all believe it; however, it is interesting how many models we come up with that are useful but ignore equilibrium.

Hierarchies of Models

A key modeling concept is that of hierarchies of models (see Figure 11.1). Very often we have models of components that describe the behavior of individual links or nodes. They describe how a traffic intersection or a highway link performs. We might have relationships that describe how fuel is consumed and how emissions are produced by a vehicle as a function of velocity and acceleration. We have models that describe how customers choose among transportation modes based on empirical data. All of these models fit together in some hierarchical way that gives some broader representation of a region like the Boston area.

This might lead to models of economic growth, of land-use, of network behavior, all of which are based on low-level, more detailed models dealing with fuel consumption, vehicle emissions, link behavior, or node behavior. Those detailed models of component behavior lead us to an appropriate macroscopic model of overall system behavior.

Modeling Issues

There are many modeling issues on which to focus. Through experience, we have learned that the following issues are important in deciding how to develop a transportation model.

Boundaries

The first issue is boundaries. From the field of structural engineering, we have the concept of free-body diagrams (see Figure 11.2). We take an element and isolate it from the rest of the structure; we analyze it by representing the entire rest of the structure as a set of forces and moments on that element—so it is a free body.

FIGURE 11.1
Hierarchy of models.

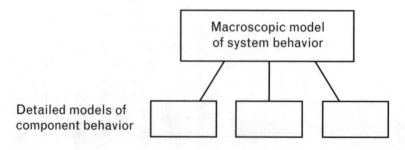

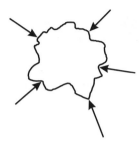

We have the same issue in transportation modeling (see Figure 11.3). What are the limits we impose in studying the system? Where do we draw the boundary? What is outside and what is inside? Are we going to consider changes in technology over time? Are we going to consider changes in land-use? Are we going to consider what our competition does, or is this model intended to tell us only what to do in the next two hours, and we assume that our competition cannot do anything in that time frame? Are we going to look at economic impacts, environmental impacts, and growth of an urban area?

Macroscopic Versus Microscopic Models

Another modeling issue is the level of detail—macroscopic versus microscopic models. For example, as we mentioned earlier, we can model a

FIGURE 11.3
A complex system—where is the boundary?

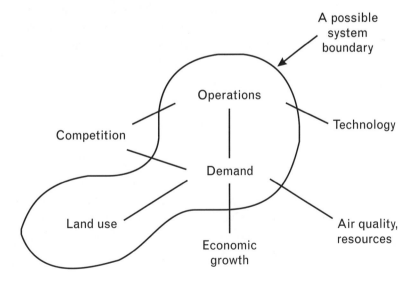

terminal as a simple deterministic delay function or through a microscopic simulation.

Static Versus Dynamic Models

We distinguish between static and dynamic models—do we use models in which we assume the key variables are independent of time or do we talk about time-dependent models that try to reflect rush-hour peaks, as opposed to steady-state operations, in our study of transportation systems? Do we assume static human behavior or do we model the reactions of people as situations change?

Stochastic Versus Deterministic Models

A fundamental question is whether to represent transportation as a stochastic or a deterministic system. We have discussed stochasticity as a characteristic of transportation, but it may well turn out in some applications that we can gain some insight and knowledge by simply representing a system as deterministic.

Linear Versus Nonlinear Models

We talk about linear versus nonlinear representations. When we talk about using linear programming to optimize a system, we are basically assuming a linear view of the world. It may be incorrect, but nonlinear models, while more correct, may be much more difficult to solve. Linear versus nonlinear models is a good example of the trade-off between constructing models that can produce answers relatively easily versus models that represent the world better but turn out to be more difficult to solve.

Continuous Versus Discrete Models

We talk about continuous versus discrete models. Speaking mathematically for a moment, we could talk about representing the world as a set of differential equations, that is, as continuous equations; or we can represent the world as a set of difference equations. At some point in our lives, we've had some function that had to be integrated—we had to get the area under the curve from $X1$ to $X2$, and for whatever reason the function was in such a form that no matter how much we looked through our table of integrals, we could not find the answer (see Figure 11.4).

What we do in that case is a numerical solution. We say, "Let's break this area up into rectangles and let's compute the area by adding the rectangular areas." We recognize that the amount of error that we will have is a function of how many rectangles we have. If we have a fairly gross representation, we may make a big mistake; if we have a detailed representation, we can come up with a more accurate answer. It is the question between continuous and discrete representations of reality that we are talking about here and how closely the discrete representation reflects the continuous representation.

The gross representation produces quick answers—there are fewer rectangles. The price you pay is accuracy. The detailed representation produces a more accurate answer but takes more time—a classic modeling trade-off.

If we have a function $y = f(x)$ that we can integrate in closed form, the way to get the area under the curve is to simply integrate. But, if we have a function that we do not know how to integrate (and this is often the case), we have to step through this in a linear fashion. If the horizontal axis is time, we are stepping the model through time.

FIGURE 11.4
Gross representation and detailed representation.

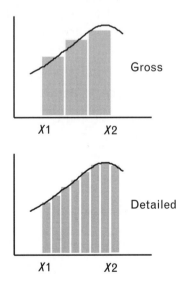

Numerical Simulation Versus Closed Form Solution

These are numerical or simulation analyses; they turn out to be very useful for generating results because we can virtually always do a numerical or simulation analysis. Sometimes, however, they can be very expensive because they take a long time to run. An analytical closed-form is better if it properly represents reality (see Figure 11.5).

We can extend this idea to stochastic systems doing probabilistic simulations by using computer programs called random-number generators to represent probabilistic behavior in the system of interest.

Behavioral Versus Aggregate Models

Another modeling question is behavioral versus aggregate representations. We have discrete-choice models for modeling what people do in making transportation and related choices—so-called behavioral models.

Aggregate representations are possible as well. The best known example is probably the gravity model, where we can model the amount of flow between two geographic points as inversely proportional to some function of the distance between them or, more generally, the resistance—distance, quality of roads, and so forth. The developers of the gravity model drew upon the insights of Newton in the context of mechanics; the idea is that perhaps cities or regions within areas work that way as well. The flow between them is inversely proportional to the square of the resistance (or some other power) between them—hence the name "gravity model." Both behavioral and aggregate models can be useful for particular applications.

Physical Versus Mathematical Models

Another distinction is between physical and mathematical models. In some areas of systems, we can physically build a model. In fluid

FIGURE 11.5 *A simulation—stepping a model through time.*

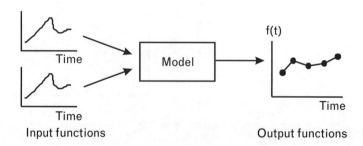

mechanics labs, we might find wave tanks; it is a physical scale representation of reality and we could do experiments using it. Models of structures are another good example.

Solution Techniques

Finally, we mention solution techniques. Getting answers from the model is fundamental to what transportation professionals do. We are often dealing with large-scale problems where we are optimizing complex systems. The brute force method of looking at every possibility is out of the question. Transportation professionals have billions and billions of options, so coming up with some efficient method for mathematically searching through decision space using optimization theory is critical.

Sometimes, scaling down the problem to make it easier to solve is an appropriate strategy when we develop models to predict performance. As noted earlier, deciding between simple representations using closed-form mathematical solutions or complex simulation models to generate answers is very important.

Understanding the System

The key is knowing what approximations we can make in representing reality, and this is where an understanding of the system, in our case the transportation system, comes into play. We know what kinds of simplifications we can make and what kinds of models we need because we understand the transportation system and the kinds of questions we want answered. We are not interested in models per se (although they are fascinating in and of themselves). We are interested in transportation; models help us to make transportation systems better.

Now, many transportation people make important modeling advances—improvements in techniques—faster optimization algorithms, superior statistical analyses, and so forth. I (and others) have observed that often those advances are achieved by people with very hard applications to consider—like transportation. To address these applications, they end up advancing the state of the art in modeling.

This is an important kind of activity. Many find it difficult—impossible, maybe—to think about improving modeling technique in the abstract—absent a specific problem. I believe that many modeling advances are made by people with intractable applications about which they need answers.

How do we decide which kinds of modeling concepts to use? To answer this, we have to go back to why we model.

Why We Model

We model to understand—insight models; we model to explain; we model to predict; and we model to improve. The fundamental question we ask when we are deciding what kind of model to use is:

What are we going to use the results for?

Be the results numerical or simply insights, how are we going use those results? That should govern the way in which we choose to model. If we are going to talk about a model that is a matter of life and death—if the model is wrong, the astronaut gets killed—we have very different requirements for the model than if we are simply looking at a model as way of developing a first-order understanding of a transportation network.

Remember:

All models are wrong.
However, some are useful.

All models are abstractions that eliminate some reality. The acid test is whether the model produces results that are useful in our area of interest.

Transportation systems are complex, dynamic, and internally
interconnected as well as interconnected with other complex
dynamic systems (e.g., the environment, the economy).

They vary in space and time (at different time scales for different components).
Service is provided on complex networks. The systems are stochastic in nature.

Human decision-makers with complex decision calculi
make choices that shape the transportation system.

Modeling the entire system is almost inconceivable. Our challenge is to choose relevant subsystems and model them appropriately for the intended purpose, mindfully reflecting the boundary effects of the unmodeled components.

If we are considering at a model to use in real time—we are actually going to control the real system in real time—we have to think about solving the model fast enough for it to be useful.

If we are going use the model for sensitivity analyses—add six trucks to the fleet, and predict level-of-service change—then we have to think about those sensitivity analyses as we build the model. We have to think about the knobs we want to turn on the model when we build the model. If we are interested in optimizing the traffic light settings in Boston, we better not represent the whole Boston traffic light grid as a single aggregate delay function. We will not get very far with that kind of representation if what we need is a microscopic representation of individual intersections.

Issues in Model Building

There are some pragmatic issues we deal with in model building. Is the model the right one? Is it "true?" We paraphrase John Dewey on the topic of how we know a model is "true."

Our model does not work in practice because it is true; rather we hold our model to be true because it works in practice.

Time and Resources

We must worry about time and resources—money, computers, and people. When does the boss want the answer? What is our budget?

Data

We have to think about data. What data do we have? What is available to calibrate this model to ensure that it is operating correctly? How expensive is it to collect more? Data is almost always a major consideration in real-world transportation systems.

Designing a Successful Model

We all want to be successful: how do we design a model that is going to be successful? By successful, we mean useful and used—at some level—by the decision-makers for whom we designed and implemented it. Success in modeling is more an art than a science. Here are some ways of measuring success.

Ease of Use

The notion of *ease of use*, developing user-friendly models, is important. Developing models that provide results in a form that is consistent with the way in which organizations make decisions is very important. If the kinds of outputs our model is producing are in some way divergent from the way in which the vice-president of operations wants to make decisions about how she runs the network, the model is not going to be used.

Convincing Models

Building models that are convincing, that make intuitive sense to the user, is important.

Growth Path

Providing models with a growth path that we can modify or expand over time as situations change is important to long-term success.

Produce Benefits

Having models that produce benefits is fundamental; we want to be able to say, "I used this model and now the MBTA runs better." To have that outcome either through the insights that management gained or through direct results that management was able to use—coming up with benefits—is very important. Success breeds success—developing a track record of useful results and demonstrable benefits is important.

Measuring Model Success

The ways in which people in practice and those in academia measure the success of models may differ substantially. As a researcher, when I look at a model, I think about concepts like unique solutions and assuring that there is a strong theoretical base.

However, when we go out to practice, people ask: "Does it help me in my job? Does it make me be a better vice-president of Marketing than I was before I had this model?" In a sense the notion is, "I do not care if you use a Ouija Board to get me these results; I do not really care about the underlying basis of this. I just want to do a better job." There is a tension there between the perspectives of the academic researcher and the perspectives of people actually using models in the field. Sometimes the difference in priorities leads to some academic models not being as useful in practice as they might be.

There are two ways that we advance in transportation. One is by advancing modeling methodologies—by thinking about how to make better models for transportation applications, where "better" may mean faster in execution, a better representation of reality, and so forth. The other is by improving transportation system performance and by developing better ways of understanding transportation systems broadly defined, often through modeling.

We advance the state of the art of methodology as well as actually improve the understanding and performance of real-world transportation systems. One approach is methodological and the other is applications-driven—both are valid.

New Developments in Models and Frameworks

So, having said all this about modeling and frameworks, what is really exciting in our current world of transportation from a modeling and frameworks perspective?

Solution of Very Large Transportation Problems

We can now solve and optimize very large transportation problems due to advances in computer technology and advances in mathematical approaches; we can address, for example, network sizes we could not consider in the past. And we have some very large networks to solve. If we look at a complex airline network and think about optimizing crew schedules, plane schedules, and level-of-service, we have a very large system to understand and optimize.

The IT Environment

The information technology (IT) environment has changed dramatically in recent years, from both a technological and cost perspective. We have object-oriented languages that allow us to write complex simulations relatively simply. We have a useful set of hardware, local area networks, and various kinds of computer platforms that allow us to use information technology in our transportation tasks much more effectively.

Real-Time Solutions

Along the same lines, computer technology has allowed us to think in terms of real-time solutions to transportation system operations problems. For example, the field of intelligent transportation systems (ITS) is based in part on our ability to do real-time solutions of complex network algorithms.

Transportation on the Agenda

Perhaps most importantly, from an excitement point-of-view, transportation is on the agenda. International competitiveness is intimately tied to the efficiency of transportation systems. Defense conversion has provided an extraordinary opportunity to convert technology and technology-oriented people to civil applications, transportation in particular. Environmental issues and, in particular, the impact of transportation systems on the environment, are major national and international issues. The institutional framework is changing—as we witness deregulation and privatization—enabling new organizational approaches. So, we clearly have some important problems to address, since transportation is in the forefront of a number of major policy debates.

What we have talked about here, in terms of models and frameworks, goes to the question of understanding the important issues in transportation. Developing and using models and frameworks to allow us to understand technology, systems behavior, and institutional factors is central to the transportation field.

We now proceed to Part II: Freight Transportation.

PART II

Freight Transportation

The Logistics System and Freight Level-of-Service

Freight

This discussion of freight begins with a simple logistics model, using it as a mechanism for introducing the important level-of-service variables. Then freight modes will be introduced: rail systems, trucking systems, international ocean shipping, intermodal systems, and air.

- Freight level-of-service—the inventory model
- Freight modes
  - Rail
  - Truck
  - Ship
  - Intermodal/international
- Summary—commonalities and differences

The Logistics Model: An Umbrella Store

A retail store sells clothing, shoes, and umbrellas and is supplied by various warehouses.

Ordering

The store orders umbrellas from the warehouse. We want them to arrive in a timely fashion, but to cost as little as possible, considering both the costs of the umbrellas themselves and transportation costs. We have to decide how often to order. Every day? Every month? Once a year? When we order, how many do we order?

There are some costs associated with ordering, which are relatively insensitive to the number of umbrellas ordered. We may order a half dozen; we may order several gross; we may order a warehouse full. The amount of paperwork will not vary substantially. There is some fixed processing cost associated with an order.

Transportation Costs

Now, consider the transportation costs associated with that order. Transportation costs of big shipments are usually less expensive per unit than small shipments. If you order one umbrella, there are shipping costs. If you order 100 umbrellas, there are some economies of scale that the transportation company has as a result of your larger shipment, and some of that is passed on to you.

Further, the price charged by the umbrella supplier is often a function of how many umbrellas you buy. If you want to buy a small number, you pay a higher unit cost for the goods than if you buy a large number. So, ordering more at a time leads to lower costs per umbrella for transportation *and* for the goods themselves.

So, we might say, "Order once a year; make a good estimate of how many umbrellas are sold per year—maybe it is 500 umbrellas—so order 500 umbrellas at once, and then we will accrue various advantages. The transportation cost per unit is cheap, the cost of the umbrellas is cheap, the processing is cheap, one piece of paperwork for the whole year. How can you fault it?" So is that what you would do? If you are the retail store operator, you would place one order for 500 umbrellas for the year? Well, probably not. Can anyone think of some reason why not?

Where are you going to store all those umbrellas?

Storage

Yes. You have this small retail store. Do you really want to take up three-quarters of your floor space with 500 umbrellas? Probably not, so storage is a factor. Anything else?

Styles change and you could get stuck.

Right. Style is very important in retail businesses. Today pink umbrellas are really a hot item. Tomorrow, nobody wants a pink umbrella. You are stuck here with 495 pink umbrellas after you sold the first 5; the supplier certainly does not want to take them back, now that they are no longer popular. There is certainly risk inherent in that.

Another issue is the cash outlay associated with this "one order per year" strategy. If you order 500 umbrellas at $20 apiece wholesale, that is $10,000. The $10,000 comes out of your money market account; instead of earning some interest on your $10,000 during that year, you are earning nothing. So you have your cash tied up in a relatively non-liquid asset—pink umbrellas. This argues against that once-per-year ordering strategy.

So, we take the opposite approach. Maybe we should order every day. With 250 business days per year, we order two umbrellas a day. Is that a good way of doing business? Probably not. The cost of ordering and transporting the umbrellas would be very high, as would the cost to you of the umbrellas themselves.

So, in practice, what you would do is balance these factors. Perhaps you would order twelve times a year. In that way, you could be sensitive to business conditions; you could be sensitive to style considerations; you would not be tying up too much of your storage space with umbrellas; you would not be incurring ordering and transportation costs that were too high; and you would not be tying up so much cash.

Deterministic Use Rate and Delivery Time

We assume for the sake of discussion, this retail outlet orders every day and the umbrellas spend one day in transit. At the beginning of each day we order and at the beginning of the next day, the umbrellas arrive. The transportation service is deterministic.

Let us further assume that the demand for umbrellas is deterministic, at a rate of four umbrellas a day. (We will relax this assumption later.) If we plot inventory of umbrellas in our retail store over time, the relationship is as shown (see Figure 12.1).

We start out with four umbrellas at t = 0. At the beginning of day zero, you order four additional umbrellas. During the day, you draw down your inventory to zero, as people purchase the four umbrellas.

The next day, bright and early, four new pink umbrellas appear on your doorstep and you go through the cycle again.

FIGURE 12.1   *In-store inventory versus time.*

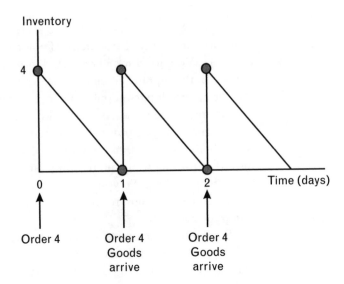

Now recognize that in addition to your inventory in the store you also have an in-transit inventory—a pipeline of umbrellas—heading from the warehouse to the store. In this example, there are always four umbrellas in the in-transit inventory (see Figure 12.2).

Longer Delivery Time

Now, suppose that your supplier of umbrellas with one-day transit time goes out of business. You find a new supplier, but he is further away and it takes a deterministic two days now for the shipments to get from this new supplier to your retail store. We continue to assume that the purchase or use rate is deterministic at four umbrellas every day. What is going to change in this business situation?

First, your transportation costs will likely be larger because the umbrellas are being shipped a longer distance. Second, since you order

FIGURE 12.2
In-transit inventory pipeline.

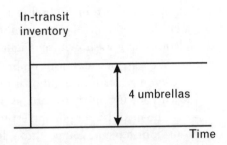

two days ahead, you now have a "pipeline" that is two days long. You still always have four umbrellas arriving each morning. However, they were umbrellas you ordered two days ago, not one day ago.

In Figure 12.3, at the beginning of each day, you have four umbrellas that are halfway to you, and you have four umbrellas that are just leaving the supplier. Your daily operation in your store is unchanged with inventory of four at the beginning of each day, drawn down to zero at the end of the day. However, in-transit inventory is larger because that pipeline is longer; in the earlier example there were four umbrellas in the pipeline at any given time, and now there are eight. Since the sum of the in-store inventory and in-transit inventory is now larger, you had to take more money out of your money market fund and convert it into umbrellas.

So, not only do you have additional transportation costs because the distance to your store is longer, but you have additional inventory costs because of the length of the pipeline.

A New, Faster Mode

Suppose a new premium transportation mode became available that allowed you to go from this new supplier to your retail outlet in one day, rather than two days. Your inventory costs would be reduced. Your pipeline is now only one day long, rather than two days long, which has value to you. And you would compare this value with the

FIGURE 12.3
Order pipeline.

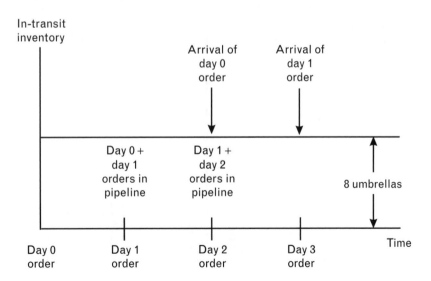

price that you were being charged for using this high-speed premium mode.

Unreliability in Delivery Time

Further, now suppose that this premium mode that takes one day rather than two is not completely trouble-free. Occasionally, on a given day, service is not provided at all on that mode (see Figure 12.4). Suppose with probability = 0.3, the system does not work at all; with probability = 0.7, it operates and it provides one-day service.

Assume that if the service does not operate on day N, on day N + 1, 8 umbrellas are shipped; if it does not go on day N + 1 either, on day N + 2, 12 umbrellas are shipped. So the flow of the umbrellas on average is 4 per day, thereby eventually satisfying all the demand at your retail store, although some customers have to wait to purchase umbrellas. The delivery time for a given shipment of 4 umbrellas is as shown in Figure 12.5.

Mathematically, $P(t = N) = (0.7)(0.3)^{N-1}$. The average value of delivery time is $(1/0.7) = 1.43$ days.

Suppose since it is a new service and the operators are still trying to work out the "bugs," they give you a deep discount on the

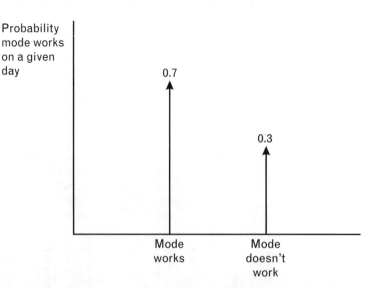

FIGURE 12.4
Unreliable transportation mode.

FIGURE 12.5
*Probability of time
until delivery.*

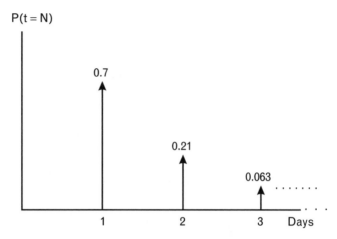

For umbrellas ordered on Day 0,
probability of arrival on a given day

transportation rate. You are deciding whether to use this new mode. You say, "I used to have this perfectly reliable (deterministic) one-day service from my old supplier. I was really happy then, but he went out of business. I then got a new supplier who gave me a nice deterministic two-day service, but in-transit inventory was very expensive. Now this new high-tech mode appears and it takes one day, only it happens 70% of the time; the other 30% of the time my goods don't arrive."

So, how would we go about thinking through whether this new service, this less reliable service, is good for us? What kinds of issues do we need to deal with in this circumstance? Under the old situations with deterministic delivery times, we could always serve all our customers on the day they wanted to be served. We have a deterministic purchase rate, and we were always sure that when we opened up at 9:00, there would be four umbrellas sitting at the doorstep that we could sell during that day. Now, there may be days on which we do not have any umbrellas to sell people.

How can we represent this? Consider the idea of back orders (see Figure 12.6). Suppose our umbrella store has very good customer relationships (or a monopoly). Perhaps our customers simply return the next day if we have no umbrellas for them when they first come to the store. So, eight arrive to buy umbrellas then. In the back-order situation, we assume the subsequent inventory can be negative. When the goods finally arrive, we are able to sell them.

FIGURE 12.6
Back orders.

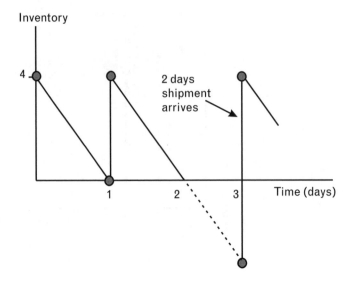

Or, possibly the customer who arrives when we have no umbrellas might say, "It says umbrellas there in big letters on your store window. I walk in to buy umbrellas, and you have none. I can't buy my umbrella here and I want it today; I am going to go down to your competitor and buy it there." It could be worse. He could say, "The umbrellas were not here; I usually buy my ties here, too, but this store doesn't have merchandise. I am not going to buy anything here anymore."

So, at the very least, we have mildly annoyed people by the fact that they have to come back a second time. Perhaps we lose that sale. They decide to buy the umbrella somewhere else. Worse, we could lose that customer forever. In the case where customers do not return the next day, eight umbrellas arrive the next day and we have a demand for only four. We have an inventory that is now too big for our demand (see Figure 12.7).

FIGURE 12.7 *No back orders.*

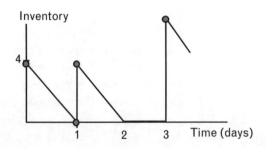

We are feeling the impact of an unreliable transportation mode. When we order a day in advance, we get our goods with a probability of only 0.7. Since this service is unreliable, we may prefer the transportation service, providing deterministic two-day service (see Figure 12.8).

It is clearly poorer service in the average value sense; we have a longer pipeline (2 days rather than 1.43 days on average). But the new service is unreliable. We may choose a longer but deterministic delivery time in favor of a shorter but unreliable delivery time.

Alternatively, we could use the unreliable service, but also keep a safety stock of, say, four umbrellas in our store to buffer against the unreliability of the transportation system—but that extra inventory costs money, too (see Figure 12.9).

Stock-Outs

The key issue is how to value a stock-out. A stock-out occurs when you run out of inventory. There is demand but there are no goods. How it is valued varies from business to business. If we have a monopoly on umbrellas in town, and won't lose the sale, we may not worry about the unreliable transportation mode. If, on the other hand, our boss, who runs a string of umbrella stores, says, "If you stock out one more time, like you did last week, you are fired," we are probably going to be stock-out averse, and will choose the mode that is going to minimize the probability of stocking out.

The impact of stock-outs varies. The operator of an assembly plant for General Motors stocks out of steering wheels; the assembly line stops for 12 hours and a high price of starting it up again is borne; that operator is very stock-out averse, so the plant has safety stock and uses a reliable transportation mode. There are other situations where the

FIGURE 12.8
Deterministic service.

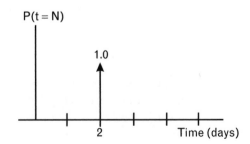

FIGURE 12.9
Safety stock.

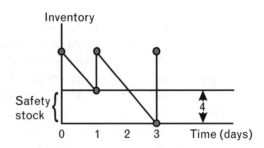

stock-out cost is less severe, but in every case, somehow, the stock-out costs must be estimated.

Service Reliability as a Level-of-Service Variable

Variability in the time for goods to travel from origin to destination is one of the prime causes of stock-outs. The term that we used for the variability of transit time is "service reliability." And that is the reliability of transportation service as perceived by a customer—in this case, the customer is the retail store.

Component Reliability Versus Service Reliability

We make a distinction between that and what we call "component reliability." Component reliability deals with concepts like mean time between failure of vehicles, which are of direct interest to the transportation operator who is concerned with the costs of running the system. Component reliability is only of indirect concern to the customer; what the customer is interested in is service reliability. Now, of course, component reliability will often have an impact on service reliability. Vehicles fail and service deteriorates. However, one could operate a transportation service so that service reliability was very good, even though component reliability might not be so good, by building redundancy into the system, in this case with extra vehicles on stand-by.

Probabilistic Use Rates

So far, we have treated use rate (or purchase rate, in this case) as deterministic at four umbrellas per day. However, in most inventory systems, use rate is probabilistic. Suppose the following figure represents the probabilistic purchase rate from this retail store (see Figure 12.10).

FIGURE 12.10
Probabilistic use rate of umbrellas.

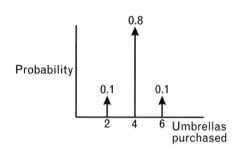

The average use rate in this particular case is four umbrellas per day, but varies from day to day. That means we could, in fact, have stock-outs, even with a perfect (i.e., deterministic) delivery system. As a result of higher-than-expected use, with six umbrellas being purchased on a particular day, we could stock out.

Consider again a one-day deterministic delivery. We order umbrellas at 9:00 in the morning and they arrive with certainty one day later. What would we have to do in this particular circumstance, if stock-outs are "forbidden," as they would be in some cases—blood supply at a hospital, for instance?

Suppose we start with six umbrellas in stock (see Figure 12.11). Since stock-outs are forbidden, we must start with six, because with probability equal to .1, we will use up all six on the first day. So how many do we have to order on that first day? Here it is 9:00 in the morning and we have initialized the system with six umbrellas in stock.

FIGURE 12.11
Inventory with probabilistic use rate.

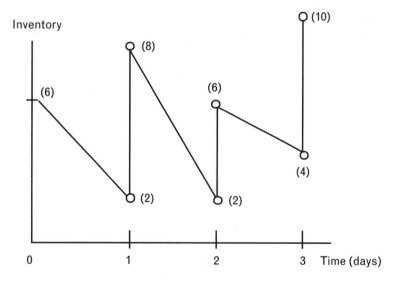

How many do we have to order, right now? (These are the umbrellas that will arrive at the beginning of day 2.)

Six.

The probability that you are really going to need six on day 2 is relatively modest. What is that probability?

One percent.

Right—.1 times .1 = .01. The only way we could stock out by ordering less than six is by using six today and using six again tomorrow. That would happen one out of a hundred times, but if "you bet your job," you would order the six.

Suppose we order the six and on the first day we sell not six, but four. So we started the day with six; we order six; we use only four on this first day. So at the end of the day we have two. Bright and early the next morning, there are the six on the doorstep, so we start day 2 with eight.

Now, we order four at the beginning of day 2 to assure we have six to start the next day, even if we use six today. Suppose we do use six today. The next morning we have the four that arrived overnight; we are at six to start the day; we cannot stock out today. To assure not stocking out tomorrow, we order six. The pattern here is clear. Once you initialize the system with six, you order whatever was used on the previous day, making the probability of stocking out equal to zero.

Now, take this example with probabilistic use rates together with probabilistic delivery times. Suppose the goods do not arrive at all on day 2. We need more inventory to buffer against that possibility. So, if we consider probabilistic use rates and probabilistic delivery times, we need to be conservative and include safety stock to avoid stocking out.

Inventory Minimization

The point is that inventory costs money. If one needs a greater amount of inventory because of unreliability in the transportation system or probabilistic use rate, you generate costs as a result of needing larger inventory to avoid stock-outs. We try to balance the costs of additional inventory with the costs of stock-outs.

....

Just-In-Time Systems

Now, in recent years, the concept of so-called "just-in-time" deliveries, often abbreviated as JIT, has been developed and refined. It was pioneered primarily in Japan and has become an important idea globally. The fundamental idea is to keep very low inventories, so as to not generate high inventory costs, by receiving goods exactly when they are needed—JIT—to keep the assembly process going, or to have goods to sell to your customers, etc. Now if one is going to operate just-in-time systems and keep costs lower by having smaller inventories (and smaller rather than larger warehouses), it requires a very reliable transportation mode. JIT with unreliability in transportation service is problematic.

There are additional complexities. In some urban areas, truck traffic is causing substantial congestion as a result of having, say, three-times-a-day deliveries rather than once-a-day to particular sites.

> Who is really paying for JIT? Someone is saving inventory costs, but at a congestion cost to others.

We always have to consider who benefits and who pays. The government and the public-at-large is paying because you need more transportation infrastructure to provide the capacity for this JIT system and congestion results for all users.

Another idea: you are shifting the burden of inventory from one organization to another. Suppose you have Toyota receiving goods from a supplier on a JIT basis. Imagine that Toyota is this supplier's best customer. It is very important they keep Toyota happy, which means they had better have goods available to deliver to Toyota. So from the supplier's main warehouse, he ships goods to Toyota several times per day because Toyota insists on just-in-time delivery. But, the supplier keeps some additional inventory in a warehouse close to Toyota in which he is carrying safety stock "just-in-case." That is the antithesis of just-in-time—just-in-case.

So the supplier may have the inventory a few miles away from Toyota just in case a bridge collapses between his main warehouse and the Toyota assembly plant. So, Toyota is running just-in-time, but the supplier is running just-in-case, with a big inventory which is expensive. Toyota's inventory is just somewhere else. It is not in Toyota's warehouse; it is in the supplier's warehouse.

Structured Inventory Model

A more structured model of inventory control is as follows: you begin at time zero with some level of inventory. The inventory is utilized (or drawn down) probabilistically.

Trigger Point Systems

We define the idea of a so-called trigger point, which is a level of safety stock (see Figure 12.12). The operating rule is: when the inventory reaches S, reorder Q items, where Q is the reorder quantity. We continue to draw down the inventory; at some point in the future, the inventory is replenished by the delivery of the goods that were ordered. Using the ideas already developed, the time from reordering to receipt is a random variable, since the transportation mode may be unreliable.

Because of probabilistic behavior, both in use rate and in the transportation system, we might stock out. We could virtually guarantee not stocking out, by choosing S very large. If you are absolutely stock-out averse, you choose a large S. Also, you use a very reliable transportation mode and are willing to pay a premium price for that transportation service.

Total Logistics Costs (TLC)

Consider now the total logistics costs (TLC) of operating the system. This includes all of the components introduced thus far, including inventory costs, ordering costs, the costs we accrue if a stock-out occurs, transportation costs—the transportation rate we are charged—and various administrative costs such as insurance, labor costs (which may vary, for example, as ordering policy changes).

FIGURE 12.12
Trigger-point inventory system.

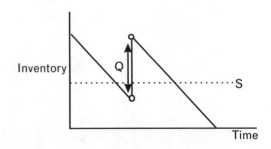

TLC depends also, and importantly, on the travel time distribution of the transportation system in use (see Figure 12.13.).

Total Logistics Costs (TLC) =
f (travel time distribution, inventory costs, stock-out costs,
ordering costs, value of commodity, transportation rate, etc.)

This probability density function defines how reliable a particular mode is; TLC is a function of the travel time distribution. As the average travel time and variance grows, larger inventories are needed.

TLC depends on inventory costs, including in-transit inventories. Keeping an inventory ties up cash on which we could be earning interest. The bigger the inventory is, the more our inventory costs are. This inventory cost depends on the value of the goods. Insurance costs would be higher as well. Also, larger inventory requires larger storage areas, which can be costly.

Also, we introduced the concept of stock-out costs. There is some probability of running out of product, which depends on the particular system under consideration.

In addition, there are ordering costs; there is some transaction cost involved with ordering. Ordering every day is more expensive than ordering every week. Finally, there is the transportation rate, which is simply the monies being paid to the carrier for providing transportation service.

We are interested in minimizing TLC. Given the travel time distribution and stock-out costs, ordering costs, the value of the goods, and the rate charged for transportation services are known, one can do the analysis that will allow you to find a combination of "Q*" and "S*" that minimizes TLC.

TLC is a function of the average travel time and variance of travel time. If everything else is held constant and the transportation mode deteriorates—higher average travel time and higher variance—TLC

FIGURE 12.13   *Travel time distribution from shipper to receiver.*

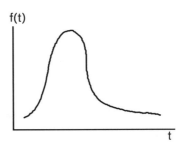

goes up because we need a larger amount of safety stock to avoid stock-out. The optimal safety stock, S*, goes up as either the mean or variance of travel time goes up. Therefore, TLC goes up if the level-of-service provided by the transportation mode deteriorates, everything else being equal (of course, often, everything is not equal—the rate charged for this poorer service may be lower, which reduces TLC), as shown in the following diagrams (see Figure 12.14).

Note that the above relationships are conceptual; they may not, in fact, be linear.

TLC and LOS of Transportation Service

So, we see here that there is a direct linkage between the quality of this transportation service and TLC. But now we ask why, as transportation people, we are interested in this analysis.

FIGURE 12.14   *TLC and transportation LOS.*

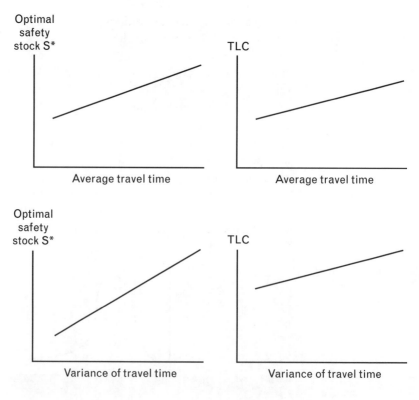

It is because from these concepts you can get a sense of what particular transportation services are worth to your customer. You can price your different transportation services, if you have an estimate of what it is worth to your customer.

Suppose a transportation company has a service that takes on average 2.5 days from A to B, with a variance of .8 days. If the company reduces the mean to 2 days and the variance to .6, this model will tell me how the customer would benefit from a total logistics cost viewpoint. If the customer was going to use this new service, he would re-optimize Q^* and S^*, because the transportation service had improved, and he would save money by needing less inventory.

The transportation company can go to the customer and say, "Here's a better transportation service. You're currently paying us $5 a ton, we are giving you service with a mean value at 2.5 days and a variance of 0.8. We will give you service for $6 a ton—raising the rate—but in exchange, you are going to have better service, a lower transit time, and lower variance." And suppose we have been able to estimate through the TLC that this better service is worth $1.50 to the customer, so he is presumably willing to pay an extra dollar a ton for that kind of service. If we can provide that service for less than an additional dollar a ton of cost to us, it is "win-win." Some people like to sit in the orchestra when they go to see *Les Misérables* and pay $80, and some people like to sit in the balcony and pay $20 to see the same show, but from a much poorer vantage point. Two different services are sold to two different kinds of theater goers. One kind of customer is paying $80 for *Les Misérables* to sit in the orchestra, and other customers are paying $20 to sit in the balcony in order to economize; both services are salable.

Market Segmentation

This is market segmentation, which is the recognition that a business has different kinds of customers who want various levels-of-service and want to pay a price commensurate with service quality. The same idea applies in transportation. The transportation carrier is not providing service only to you, the umbrella retailer, but to the Toyota assembly plant and to a coal-burning power plant as well. The transportation company provides different services to all these businesses using the same infrastructure. Those services are, in some cases, of very high quality. High rates are charged for them; the transit time is fast;

the variance of those transit times is low. But unlike the theatre example—where the cost of providing a high-priced seat is the same as for a low-priced seat—the costs to the transportation company of providing this high-quality service is usually high.

On the other hand, there is a set of services that are of poorer quality. Low rates are charged for them. The transit times tend to be long, and the variances tend to be high; but they are of lower cost for the transportation company to provide. There are customers that prefer the high-quality, high-price service, and those that prefer low-quality, low-price service. A railroad company or trucking company has various kinds of services, and will seek different customers who have different perspectives on logistics costs to use these services. The person who is running a coal-fired power plant may have a very different inventory objective and a very different cost structure than our umbrella retailer, so we will find that there will be different transportation services that will be attractive to different customers.

Allocating Scarce Capacity

Railroads need to allocate capacity (e.g., train capacity) among various customers with very different service requirements. Capacity is allocated among customers who require their high-quality service, for which they are willing to pay top dollar, and low-quality service for customers who do not want to pay so much.

From a carrier viewpoint, the idea is to make a profit in each service class. If you are a railroad company who is providing high-quality service at a high rate, that service is probably high-cost to you. Contrast that with low-quality service for customers who will be charged less; the presumption is it costs you less money to provide that service. The idea is to assure the costs and revenues in each of those classes are such that you are making money on each.

Yield Management

And, further, you should try to allocate capacity among service classes so that profits are maximized. This step is called "yield management." We will discuss this idea further in Chapter 29, Intercity Traveler Transportation: Air.

Costs

As discussed in Key Point 13, computing the costs associated with various services can be a very difficult thing to do. For example, suppose one has a rail right-of-way with three kinds of services: high-quality, medium-quality, and low-quality service. The costs of maintaining that right-of-way are part of the cost structure of your company, but how do you allocate those maintenance costs among those various kinds of service? You have to make some kind of rational judgment about how to allocate so-called joint costs among the various services to make a judgment about what to charge for that service in order to make money.

There is a continuing debate between railroads and trucking companies about the way in which costs ought to be allocated in transportation systems. The railroads make an argument that the trucking industry is not paying enough of the costs of the publicly provided highways, and that there is a cross-subsidy from individual car drivers to the trucking industry, because the former pays to maintain the latter's right-of-way. The railroads would like to see the truckers pay a larger fraction of the maintenance costs of highway infrastructure, so that their costs would increase and permit the railroads to be more competitive on price. The truckers, of course, see the cost allocation differently.

Other LOS Variables

We have talked thus far about three parameters on which shippers would evaluate freight service and make a modal decision. We talked about travel time—an average value measure—service reliability, which is a variability in travel time, and rate—the price that the carrier charges the shipper. But there are a number of other LOS variables as well.

Loss and Damage

Loss and damage is a level-of-service variable. If you are delivering finished automobiles from an assembly plant to the car dealership, you want to make sure they get there without a broken windshield or a stolen side-view mirror; loss and damage for that kind of high-value freight is important. (In the railroad industry, loss and damage runs between one and two percent of gross revenues.)

Rate Structure

Rate structure is another level-of-service variable. For example, to what extent can shippers negotiate discounts in exchange for long-term commitments to a carrier?

Service Frequency

Service frequency is another level-of-service variable. In ocean shipping, once or twice a week frequency is not uncommon. So a competitor coming into the market with three- or four-times-a-week service could compete very vigorously.

Service Availability

Availability of service as a function of location is an important parameter in shipper modal choice. If you do not have a rail siding at your plant, using rail service can be more difficult, perhaps even infeasible. To use rail service you have to transport your goods by truck to an access point on the railroad network. This is one of the inherent advantages of truck over rail. The highway network being as universal and ubiquitous as it is, provides service virtually between any pair of points; the rail industry, with a fixed dedicated network provides direct service to a much smaller number of points. This has led to rail-truck partnerships in providing origin–destination service.

Equipment Availability and Suitability

Availability of equipment is an important level-of-service variable. For example, in order to use the rail system, the customer needs freight cars. A freight car has to be delivered to the plant so it can be loaded. So, we have to be concerned about the time it takes to deliver empty equipment to the customer (shipper) for use on the network.

Suitability of equipment is also a critical issue. For example, refrigerated trucks and refrigerated freight cars are used for the distribution of produce. If a commodity is perishable and the rail industry or the trucking industry cannot provide you with a working refrigerated car, then the fact that they can give you a boxcar or a conventional truck does not help you very much. Cleanliness of equipment can be important. If the railroad provides a hopper car that was just used for some chemicals without cleaning it, the car is not suitable for transporting grain.

Shipment Size

Shipment size is a variable that is very important to shippers in deciding what modes to use. Different modes can handle shipments of various sizes. The trucking industry, for example, is not particularly efficient in transporting bulk commodities on a large scale, like coal and grain; on the other hand, the railroad industry and the ocean container-ship industry are good at that.

Information

The availability of accurate information about the shipment—where it is and when it will arrive—is a level-of-service variable. If the customer knows a shipment is going to be delayed, he can make tactical moves that will help. The customer (or the railroad!) can airfreight those steering wheels to the assembly plant at a premium price to avoid shutting down, if he knows that the steering wheels that were coming by rail are not going to make it on time. Having information allows one to make more effective use of the transportation system. In general, one will be more likely to select a mode that can provide better real-time information about shipments. The idea of "in-transit visibility," which provides continuous real-time location of shipments, was pioneered for military applications and is now commercially available.

Flexibility

Finally, we consider flexibility as a LOS variable. Is the transportation company one that says, "This is the service. Take it or leave it," or is the transportation company one that accommodates the particular needs of shippers? Shippers need flexibility as their needs change, often very dynamically and often in very unexpected ways; transportation companies that are flexible in accommodating those changes in, say, origin-destination patterns, will be more successful.

   Providing flexibility often costs the carrier money. We introduced this idea in the elevator example; we discussed the three shafts all accessing the sixtieth floor, to provide flexibility, even if we were currently going to provide service from floors zero through 20, 20 through 40, and 40 through sixtieth with each of the three elevators, respectively. Building the shafts all the way to the 60th floor for all three elevators is flexible. It allows us to operate that transportation system more effectively if demand patterns change. It is also a more expensive transportation system to build, so flexibility valued by shippers may be costly, because the carriers incur cost by having flexibility inherent in their operation. These costs are passed on to the shippers. If a shipper has an operation in which flexibility is important, that shipper will pay some premium for flexible service, which may well make sense, considering TLC.

CHAPTER 13

Railroads: Introductory Concepts

Modes

We now move on to transportation modes and how they operate and deliver different levels-of-service depending on investment and operating decisions.

Railroads

We start with a discussion of railroads for two reasons. First, it is an important freight mode in the United States, carrying about 36% of the ton-miles. Second, it is a good illustrative mode. We can use it to introduce concepts that are relevant to other modes as well. We quote from a book published in London in 1833 [1]. The author, Mr. Richard Badnall, says, "It is, I believe, universally acknowledged that in all countries the rise of prosperity mainly depends upon the convenience of conveyance from place to place." And then, further, goes on to say, "No nation can promote its real interests more effectually than by encouraging, in every possible way, the establishment of good roads rapid and convenient modes of traveling; for according to such convenience will be the equality and price and the abundance of the supply of produce, the real value of land and another property, and as before stated, the increase of wealth and comfort among all orders of society." (See Key Point 27).

Those quotes are over 160 years old. The idea of building transportation infrastructure as a means for encouraging economic development has been around for a long time, as has the technology of railroading. Indeed, railroads began operations in England in the 1830s.

Rail Technology: A Basic View

Modern railroads are based on the technology of steel-wheel on steel-rail. Power is provided by locomotives; diesel and electrical locomotives are in common usage. Technologies have, of course, changed dramatically over the years. Steam technology was the primary mode for many years. Diesel is currently the major power mode for freight railroads in the U.S. On some portions of the infrastructure, electric power is used, usually for passenger rather than for freight transportation. We will not dwell upon the technology characteristics of the railroads.[1] Here, we will discuss rail at the systems level.

Low-Cost Transportation

Rail is fundamentally different in operation from a highway. Fixed rails provide guidance and control. There are traction characteristics in steel-wheel on steel-rail that differ greatly from rubber tire on concrete or asphalt. The fundamental railroad philosophy has been to spend money on a specialized right-of-way limited to particular kinds of vehicles: locomotives and freight and passenger cars. The concept is, by developing this high-cost, specialized right-of-way, we gain tremendous operating advantage in our ability to haul freight, often bulk commodities like coal and grain, at reasonable speed, safely and at low cost. That is the philosophical underpinning of railroading. However, by virtue of having a specialized infrastructure, the network is not universal—as are highways. There are some places the rail network simply does not go. Access to the network is essential if you want to use rail services.

It is worth commenting that in the United States, the freight railroad system is probably the best in the world. On the other hand, the U.S. passenger railroad system is probably among the worst in the developed world. The public image of railroads is shaped by their passenger services. Therefore, they are not held in high regard here in the United States; however, U.S. freight railroads are well-respected by countries throughout the world.

---

1   A suggested reference for those interested in understanding the technological concepts behind how railroads operate: Armstrong, John H., *The Railroad: What It Is, What It Does*, Omaha, NE: Simmons-Boardman Books, Inc., 1993.

Railroads as a Monopoly

The development of railroad technology in the nineteenth century preceded the development of the truck/highway system for freight by decades. In the early history of railroading in this country through the turn of the century, the competition from roads was modest indeed. The quality of the roads in the United States, and internationally, was not high; therefore, railroads had what amounted to a collective monopoly in the development of freight markets in the United States. Of course, there was competition among railroads, but rail had no modal land competition. This was true not only for bulk commodities, but essentially for all commodities; for many years railroads operated as virtually a collective monopoly in the transportation of goods around the United States.

Regulation and Deregulation

Some of the woes of the railroad industry in more recent years, it could be argued, stem from those early successes. They were so successful in dominating the marketplace, that the shipping public—the farmers in particular—rose up to ask the government to regulate the monopoly powers of the railroad industry. In 1887—more than a century ago—the Interstate Commerce Commission (ICC) was formed, with the mission of properly regulating this monopoly operation.

Today the railroads are no longer a monopoly—far from it. They have a great deal of competition from other land modes, the highways and pipelines in particular. The regulatory apparatus that was put in place over a century ago to curb the excesses of monopoly behavior by the railroads has been disassembled much more slowly than the market power of railroads deteriorated. In 1920, the railroads carried about 85% of the ton-miles in the United States; in 1970, that market share had been reduced to about 36%—that is a rather substantial change—a virtual monopoly at 85% to a little over a third of the market over that 50-year period. If one considers the share of revenue rather than the ton-miles, the shift would be even more dramatic.

However, in the last two decades a good deal of deregulation has occurred; this is a relaxation of the regulation by the federal government of the rail industry with regard to pricing, abandonment of services, mergers, treatment of captive shippers, etc. The rail industry would argue that they are still regulated out of proportion to their true power in the marketplace. They would say that their monopoly has long since disappeared and that they should have much more freedom of action. Certainly they do have much more freedom of action than they did prior to 1980, when the Staggers Act was enacted; this legislation changed the playing field in freight transportation in a very substantial way. It can be argued that the current success of the U.S. railroad industry stems in large part from the deregulation wave of the 1980s. Many feel this deregulation was long overdue and that the competition from the trucking industry, accelerated by the development of the Interstate System beginning in the 1950s, did substantial damage to the U.S. railroads, financially and as freight-carrying enterprises.

Railroad Management

Railroads today are viewed as relatively traditional in their approach to the world of commerce. However, for those interested in the development of management ideas, if you consider the period from 1870 to the turn of the century, and read some of the works by contemporary historians, such as Professor Alfred Chandler at Harvard, the railroads were true management innovators [2]. Railroad companies were enormous enterprises by late nineteenth century standards. They were geographically distributed, unlike manufacturing organizations that tended to be highly concentrated. The management structures that were required for these complex organizations to be effectively run were, at that point in history, extremely innovative. In fact, one could argue that the management structures of American companies, and companies around the world, closely followed the hierarchical model of railroad management for many years.

In more recent years, the advantages of the hierarchical command-and-control structure that these railroads developed—a virtually paramilitary operation—became a disadvantage in the more nimble, information-rich, communication-rich society in which we now live.

High Fixed Costs

The railroad industry is a high fixed-cost industry. For ton-miles or some other measure of production—a ton-mile is simply one ton transported one mile—the relationship is conceptually shown in Figure 13.1.

There are substantial costs before you transport anything. You first need to build all this complex and expensive infrastructure. But once

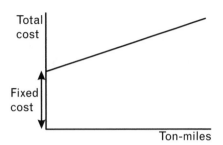

FIGURE 13.1 *Railroad cost function.*

you have that infrastructure in place, you have relatively modest variable costs for operating the system. It is obviously in your interest to attract high traffic volumes, because the average cost per ton-mile will go down substantially as ton-miles go up, because you have more ton-miles over which to spread your high fixed cost (see Figure 13.2).

U.S. Railroads Own Their Infrastructure

Recognize that the freight railroads in the United States own the infrastructure. They own the right-of-way, the track, the ballast and the ties. They own the entire infrastructure and they need to maintain it in order to operate their system. That maintenance is, to a certain extent, regulated by the federal government through safety standards. But, simply put, their high fixed costs derive from the fact that they paid for and own the infrastructure, and continue to pay for maintaining it.

Didn't the U.S. federal government provide subsidies for the original construction of the railroads through land grants?

Yes. At the time of the initial construction of the railroads, substantial amounts of land and money were provided by the federal government to allow entrepreneurs to build these railroads through virgin

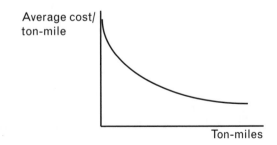

FIGURE 13.2 *Railroad average cost function.*

territory where there were not many people or much economic activity. What drove this was a national vision of a universal and transcontinental railroad system. Subsidies and real estate were granted to these railroads as the network was put in place; in fact, the wealth of some of these railroads results as much from the development of that real estate as it has from the operation of the system. Some of the great railroad fortunes amassed in the 19th and early 20th centuries resulted from profits on the construction itself. However, now the ownership costs of that right-of-way impinge on the railroads, leading to a high fixed-cost operation.

Contrast railroads with the trucking industry. The latter uses the highway system as their infrastructure, which is publicly owned. The question of how much of that infrastructure trucks ought to pay for, through their road-use taxes and fuel taxes, is a matter of debate between the trucking industry and the railroad industry. The railroad people feel the trucks grossly underpay for the wear and tear they cause on the highways; the trucking people feel that they grossly overpay for the wear and tear that they cause on the highways; the truth, as always, is probably somewhere between these extremes. A key difference in the modes is in ownership of infrastructure. If, in the limit, a trucking company does not use the infrastructure at all, it does not pay for it at all (or modestly). The railroads pay substantially, even if they use their own infrastructure at a modest level.

Here is an interesting question to think about from the point of view of U.S. consumers. Let us suppose road taxes were raised for the trucking industry, having them "pay a fair share for their use of the highways." It would be interesting to do an analysis of how those costs would be distributed. The trucking industry will be passing on some of those costs to the people that buy transportation services from them, like food companies, for example, shipping to supermarkets. The food companies will, in turn, pass the increases on to the supermarkets, which will pass them along to the general public. The issue of how those costs are distributed is quite complex and really beyond the scope of what we will talk about here, but interesting nonetheless.

Our discussion has suggested differences between the cost structure of rail and truck. This difference between rail and truck cost functions is conceptually illustrated in Figure 13.3.

Note that we make no claim that the costs are linear. We simply illustrate that trucking is a high-variable-cost, low-fixed-cost business; railroads are a high-fixed-cost, low-variable-cost business.

FIGURE 13.3
*Rail versus truck cost
functions.*

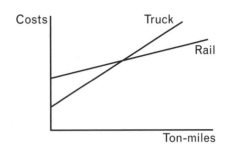

Freight Transportation Statistics

Next, we examine modal shares of various carriers. The following data
comes from a very useful book, *Transportation Statistics Annual Report*,
that is published by the Bureau of Transportation Statistics, an agency in
the United States Department of Transportation (DOT) [3].

Modal Shares

We present U.S. modal share in tons, ton–miles and revenues. Tons
simply refer to the numbers of tons carried, without reference to how
far they are carried. Ton-miles are a combined measure, incorporating
the weight of the goods and the distance they are transported; revenues
are simply the prices paid by shippers for transportation services. The
data is shown in Table 13.1.

Modal Share: Tons

Look first at tons. Rail has a 26.8% share and truck 41.5%. The pipeline
share, at 15.8%, often surprises people. People see a 0.1% tons modal

TABLE 13.1   United States Modal shares, 1994.

	TONS (%)	TON-MILES (%)	REV (%)
Rail	26.8	34.8	8.8
Truck	41.5	24.5	78.7
Water	15.8	21.9	6.0
Pipeline	15.8	18.5	2.5
Air	0.1	0.3	4.0

share for air and discount its importance, but we will soon see that is not the whole story.

Modal Share: Ton-Miles

Look now from a ton-mile point of view. Rail, on a ton-mile basis, jumps to a 35% share. Trucking, on that basis, drops to 24.5%; water goes up to 21.9%; pipeline does not change very much (18.5%); and air transportation is 0.3%. So we can draw some conclusions about lengths of hauls for these various modes. Railroads carry their freight further than trucks. Also, air certainly is not a short-haul mode. Its modal share goes up by a factor of three when one shifts from tons to ton-miles, so people are using air freight to ship goods substantial distances.

Modal Share: Revenues

Look now at revenues. We look at the nation's freight bill, and divide it up in terms of the dollars each mode attracted. Railroads have an 8.8% share of the revenues. They carry 27% of the tons; they carry 35% of the ton-miles; they receive about 9% of the dollars. The conclusion is that they are charging low rates for transportation of low-value goods where transportation rate is more important than service quality. Presumably they are providing lower-quality service. Trucks go up to 78.7%; trucks are carrying high-value goods, providing a high quality of service and charging higher rates for it.

Water generates only 6% of the revenue, and the same conclusion can be made for water as was made for rail. What freight gets shipped by inland barge? Coal, sand, and aggregate for highway construction. Those barges move slowly. And they are charging very low rates; it is a low-cost, low-revenue, low-service-quality business.

We again emphasize that when we comment on low-service quality, we are not saying that in a pejorative way. It is not any more pejorative than *Les Misérables* selling balcony seats at a cheaper rate than orchestra seats. There is a market for balcony seats, just as there is a market for low-quality transportation service.

Pipeline generates only 2.5% of the revenues, despite carrying 18.5% of ton-miles, again, a very low-cost mode. Finally, we noted air with 0.1% share of the tons, but they received 4% of the revenues! They are charging a premium price for a premium service to a limited market. This is a market that says, "My gosh, we just ran out of steering wheels.

We need those here by tomorrow morning, no matter what it costs, or we have to shut down the assembly line. We will pay whatever they want to charge us for this service."

Different Modes—Different Roles

So there is a place for each of these modes in the service spectrum.

Railroads Continued

Now we discuss rail and the various services it provides.

Commodities

Coal represents 40% of the tons carried by railroads in the United States. While coal is the backbone of the rail business, it is a low-value backbone, not requiring high level-of-service with high revenue/ton or ton-mile. The second-largest commodity is farm products at about 11%: largely grain and corn—again bulk commodities. Third is chemical products at 9.4%—this is more high-value, but certainly bulk in nature and still a low-value commodity as compared with manufactured goods.

Freight Car Types

Railroads transport goods in freight cars. In that general category is a variety of equipment types. In Morlok's text [4], he has a succinct and useful description of each of the general equipment types, along with some photographs of them. These various types of cars reflect different shipper needs. First, is the plain vanilla "boxcar." This is simply a box with a door. It is an enclosed vehicle, which gets your freight out of the weather.

Second is a "flat car." A flat car is precisely that. It is simply a set of wheels with a platform. Occasionally, they will have a front and a back, but not sides; they are often used for containers—simply a box made of aluminum or steel, in which goods are stored—you can load one or two of those appropriately fastened down on a flat car. We also put truck trailers on flat cars. These have wheels, just like a trailer connected to a tractor; for intermodal transportation, one can simply take that trailer, unload it from the flat car, and attach it to a tractor to operate on the highway. A container on a flat car obviously cannot be transported that way since it has no wheels. In that case, you need a truck with a flat bed on which to place the container.

Double-Stack

In recent years, there has been an innovation called double-stack container service in which containers are stacked two high (see Figure 13.4). Thus, the flat car can carry four containers rather than two.

Through double-stacks, you can change your productivity by close to a factor of two, with modest impact on overall costs. This makes rail more competitive with truck and other modes. If it costs the railroad less, it can charge less. It makes rail look better to that inventory manager who is trying to minimize logistics costs.

Of course, there is a needed investment in equipment to handle double-stacks and wheel-loadings on the track will be higher, implying higher maintenance costs. Bridge and tunnel clearances are an issue on some parts of the network. But even considering that, double-stack is a strong competitive weapon for the railroads.

Double-stack trains are now a very important part of the railroad business. In fact, intermodal transportation for the rail industry is the fastest growing part of the railroad business. The overall revenues derived from intermodal transportation by railroads—transporting containers on flat cars or trailers on flat cars—is probably now in excess of 10% of the gross revenues, fueled in large part by the shipments from the Pacific Rim to the West Coast of the United States. The emerging manufacturing organizations in the Pacific Rim nations are shipping goods into the United States. Double-stack arose as an innovation, partially in response to these flows.

FIGURE 13.4
*Conventional flat car
(top) and double-stack
(bottom).*

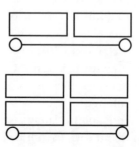

Gondola Cars

Other kinds of equipment include gondola cars; these are often shaped as shown (see Figure 13.5).

They have no roof. They do not protect your freight from the elements; they are used for such items as grain and coal that are not especially perishable. Gondolas are rough-and-ready transportation—inexpensive transportation to accommodate a particular part of the market. A "cousin" to gondola cars are hopper cars. Hopper cars are shaped basically like gondola cars. The fundamental difference is they have a provision for opening the bottom, and therefore, allowing off-loading, say of grain, by a gravity feed. Hopper cars can be covered or uncovered.

Tank Cars

Another equipment type is tank cars. They are cylindrically shaped. Chemical companies, for example, ship liquid chemicals in tank cars. There is always concern for damage in those kinds of cars, which are often carrying chemicals which, if spilled into a river or on the roadside when such a car is in a crash or derailment, can be quite hazardous.

Hazardous Materials

The industry has to be concerned with clean-ups; liability associated with these hazardous spills are potentially high. If such an accident happens in an urban area it could be an extremely serious "Bhopal, India" occurrence, if not properly handled. There has been a lot of work over the years in assuring that these cars are properly labeled so that the clean-up crews know what they are dealing with; knowing simply what is in the spill is fundamental to safely handling it. Further, there have been important physical design improvements to tank cars.

FIGURE 13.5
Gondola car.

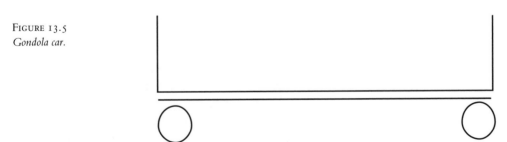

Refrigerator Cars

Refrigerator cars are specialized equipment. The term used in the industry is "reefers." They are used to transport, for example, fresh produce from the fields in California to points east. They have to operate reliably because, if the cooling unit goes out of service, the loss and damage is total. Lettuce that has been sitting out in the desert without proper refrigeration is a complete loss.

Auto-Rack Cars

There is specialized equipment for transportation of finished automobiles—auto-rack cars. Often they are fully enclosed in order to permit more secure operation—avoiding damage to high-value finished autos.

Specialization in Freight Car Types

We have seen a trend over the last 20 or 30 years toward more highly specialized freight cars in the railroad industry. The automobile industry is a big customer of the railroad industry. They insisted on the development of enclosed auto-rack cars and to nobody's great surprise, the rail industry has accommodated that command. They are an important customer; just as our umbrella store is going to accommodate the demands of an important customer, so will the railroad industry.

We have seen a steady growth in specialized equipment that responds to shipper needs. That is a way of improving the level-of-service—by giving shippers the kinds of equipment they want. In the 1920s, when rail had a virtual monopoly, the rail industry did not have to exhibit that kind of customer service. The competition was not there. If you wanted to use railroads, and many people simply had to, you used the equipment they had and you lived with it. In this more competitive age, the railroads cannot succeed with that kind of strategy because there is a lot of intermodal and intramodal competition. So the rail industry improves their level-of-service with more specialized, more costly equipment.

This specialization has implications on operating practices because of the need to more intensively utilize a resource that costs a great deal, like an enclosed tri-level auto-rack car, than to utilize a resource that costs much less like a plain vanilla boxcar. Common sense says that you would really want to keep that large investment moving, where if that small investment sits for a day or two, it is less critical. So, operating policies should be changed in response to that change in the cost structure; some railroads have made those changes more effectively than others.

....

Railroad Growth and Rationalization

We continue by introducing the term *rationalization*, often associated with the rail industry, although it could be associated with any transportation system. It deals with shrinking the capacity of a transportation system because of lower traffic levels or changing geographical distribution of traffic. This could be accomplished by shrinking the physical network, the size of the work force, the rolling stock, and so forth. Here, we examine the number of miles of right-of-way in the U.S. rail system and note how it has changed over the years (see Table 13.1). The industry grew to accommodate a growing United States, and then was rationalized in response to overbuilding and competition.

In 1840, there were 3,000 miles of rail trackage in the United States. By 1860, one year before the Civil War, there were 30,000 miles—a factor of 10 growth in 20 years. In 1880, there were 93,000 miles—a factor of about 3 growth in that 20-year period. And, from 1880 to 1900, growth to 193,000 miles took place, which is a factor of about 2 over that 20-year period. The growth rate then slows up substantially. In

TABLE 13.1   U.S. rail track mileage (*Source:* Railroad Facts, 1993 ed., Association of American Railroads, Washington, D.C., 1993.)

YEAR	MILES (000)	YEAR	MILES (000)
1840	3	1930	260
1850	9	1940	230
1860	30	1950	223
1870	53	1960	217
1880	93	1970	206
1890	167	1980	164
1900	193	1985	145
1910	241	1990	120
1920	260	1992	113

1920, we have no trucking industry to speak of in the United States Most roads are not paved. We have close to a monopoly system for freight railroads, the main competitors being inland waterways at that time. Certainly, air freight was the most modest of competitors at that point, possibly beginning to develop some United States mail business.

From 1900 to 1920 the network grew to 260,000 miles, a factor of 1.25 in that 20-year period. By 1940, some modest shrinkage in the system begins—from 260,000 to 230,000 miles. In 1935, the Motor Carrier Act of 1935 was signed. This was landmark legislation regulating the trucking industry and beginning the major period of competition between truck and rail. So we see a shrinkage of about 10% in rail trackage. We go forward to 1960. By this time, we have finished World War II and the Korean War. We again see a modest dip, from 230,000 to 217,000 miles.

Major Shrinkage of Rail Network

Then we really begin to see some major rationalization. In 1980, the network has shrunk to 164,000 miles. The size of the network is about equal to that in 1890; the system was rationalized—shrunk to reflect the strong competition of the trucking industry and the overbuilding of rail itself. By the year 1990, the industry is 120,000 miles. The industry is well below half its peak mileage attained in 1920 and the trend continues. In 1992, about 113,000 miles were in operation.

Substantial Growth/Substantial Rationalization

So you see an industry that had gone through a substantial growth followed by a substantial rationalization. It grew very quickly during a period where it had virtually a monopoly on freight transportation in the United States and the United States was growing rapidly. Indeed, tremendous overbuilding occurred in the latter part of the 19th century and the early 20th century as entrepreneurs occupied—and overoccupied—every possible geographic and market niche. With fortunes being made on the subsidized construction of railroads, this is not surprising. Then, as both intermodal and intramodal competition developed, the network shrank along with shrinkages in the size of the labor force, until it is down to a size at which the railroad industry is economically viable. They can successfully operate at a profitable level.

Mergers

Now, several factors enabled this shrinkage. One was a series of major rail mergers that led to, in many cases, a dramatic shrinkage in the number of track miles in the United States. You have two parallel railroads that had been built during the boom time. Now, there is competition with trucking. There is not enough traffic to really keep both railroads alive. The two railroads merge. Trackage is cut substantially, and that single railroad can profitably operate.

Branch Line Abandonment

Also, the railroads abandoned branch lines—relatively lightly used branch lines serving areas where manufacturing facilities had left or lowered their output substantially. So, the industry rationalized their network by eliminating unprofitable branch lines.

Both these actions—mergers and branch line abandonments—have accelerated in recent years as the regulatory posture of the U.S. government toward the railroad industry has been relaxed.

Guarding Against Monopoly Power

The Interstate Commerce Commission (ICC), was formed by the Interstate Commerce Act of 1887, in a period in which railroads had a monopoly. (On January 1, 1996, the ICC ceased to exist as such, although many functions were absorbed into the Surface Transportation Board—or "Surfboard"—in the U.S. Department of Transportation.) The ICC worried, over the years, about protecting the general public from monopoly behavior on the part of railroads. They have been concerned with the implications of mergers on competition.

The United States has anti-trust laws that protect against monopoly power. Suppose two railroads merge where they have been fierce competitors in a particular corridor. The United States government is concerned about where that leaves shippers who now do not have the benefits of two railroads competing with each other for their traffic, thereby serving as a brake on rising rates; rather, they now have only one railroad competing with perhaps trucking. So, the government was very cautious about allowing mergers to take place in view of the anti-competitive aspects, although in recent years we have seen a substantial relaxation of that constraint.

Also, on the question of branch line abandonments, the ICC had taken the position that protecting those factories that were served by a branch line was important; often, they would create some conditions that were difficult to fulfill for a railroad to meet before they would permit them to abandon a rail line. A shipper could argue, "I built my factory here because there was a branch line and now, 10 years later, just because somebody else up the road has gone out of business, the railroad wants to abandon this branch line, leaving me high and dry. I have a real problem." The Commission has generally been sympathetic to those problems. Over recent years, again, there has been a greater willingness on the part of the government to allow railroads to rationalize their physical plant, when decreased traffic made that reasonable.

Cross-Subsidies

When the physical plant is not rationalized, realize that there are cross-subsidies. This is a continuing theme in transportation. There are subsidies that go from user to user, from mode to mode, from sector to sector, from geographical area to geographical area. If indeed a railroad company is required to keep a particularly low-profitability or even a money-losing branch line open, someone is going to pay for that. The railroad company will try to recover that cost; if within the marketplace they are able to lay that cost off onto somebody else, they will do so. Maybe the coal companies, who are captive to the rail industry (since transporting large quantities of coal by truck is not feasible), absorb those costs.

That concludes our introduction to railroads. We continue by discussing railroad operations in some detail.

REFERENCES

1. Badnall, R., *A Treatise on Railway Improvements*, London, England: Sherwood, Gilbert and Piper, 1833.

2. Chandler, A. D., Jr., *Strategy and Structure: Chapters in the History of the Industrial Enterprise,* Cambridge, MA: The MIT Press, 1962.

3. *Transportation Statistics Annual Report 1994*, Bureau of Transportation Statistics, U.S. Department of Transportation, January 1994.

4. Morlok, E. K., *Introduction to Transportation Engineering and Planning*, New York: McGraw-Hill, 1978.

CHAPTER 14

Railroad Operations

....

Railroad Operations

To introduce railroad operations, consider a system in which we have a shipper of some particular goods and a receiver of those goods and a rail network with various terminals through which that shipment will travel to get from shipper to receiver (see Figure 14.1).

The process is as follows: we begin with a placement of empty cars at the shipper's siding. The shipper would call up the railroad and say, "I need three freight cars." These freight cars would be dropped off by a local train at his siding for loading. The shipper completes the loading and would communicate to the railroad that they could be picked up; those three cars would be picked up by a local train. This would take place on a branch line of the system; a local train would come out from a main terminal (terminal A), pick up those three cars,

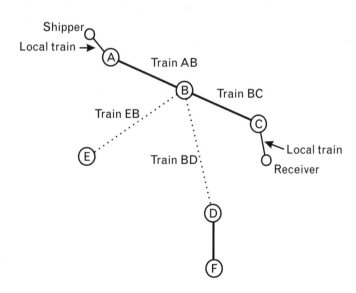

FIGURE 14.1 *Path from shipper to receiver.*

presumably pick up cars from other shippers along that same line, and bring them to terminal A. Depending on traffic levels, the local train might provide daily service, or perhaps less—four times per week, for example.

These shipments can now be placed in a "road train" that will go between two major terminals on the system, terminal A and terminal B. Now, train AB—so designated here because it goes from point A to point B—would have various "blocks" of traffic intended to go between various points on the network. So, traffic that was destined for terminal B, for C, for D, and for E would be on train AB. So on this train we have traffic destined for four different points on the network—direct service to B and service to terminals south and east—C, D, and E. Figure 14.2 shows the various trains on the system with traffic destined for various points.

Blocking

We emphasize that, although we show train AB with various blocks, the freight cars may not physically be in that order—that is, they may be unsorted. There will be trains destined for various points on the network and whether they have been blocked at terminal A into contiguous sets of cars is an operating policy question. It may make sense to do that blocking at terminal A or it may make sense to do that blocking at terminal B; where the sorting out is done is a critical question in rail system operation. So that sketch of train AB is conceptual. It does not reflect physical contiguity of cars on the train.

For example, in Figure 14.2, we show two possible configurations of train BD—the first has D and F cars randomly scattered in the train. The second has D and F cars blocked into contiguous sets. Presumably in this second case, it is more efficient to do this work at B than at D.

FIGURE 14.2   *Blocking patterns.*

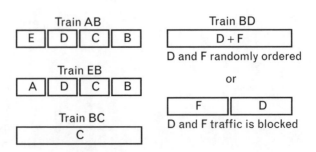

Now, in addition to train AB, we have a road train EB that would have various blocks destined for A, B, C, and D; this train would have blocks that would be destined for B as an immediate destination, as well as blocks going on to A, C, D, and F.

Consolidation

The fundamental operation in railroading is consolidating freight cars destined for the same destination into trains. Basically what we are trying to do in this system—a high fixed-cost system—is take traffic from E and A destined for C and block it into a single set of cars that will go together from B to C at presumably lower cost than in the case of A-C and E-C traffic going separately. We want to consolidate.

Now, inherent in this operating mode is a complicated terminal operation. Later, we will discuss terminals in more detail. But for now, we are simply introducing the idea of sorting freight cars physically, moving them from inbound train to outbound train.

Missed Connections

With the complexities inherent in railroad terminals, you do not always move cars onto the appropriate outbound train in a timely fashion—cars that were intended to go from E to C via B might, for example, miss the outbound connection going from B to C and be delayed as a result. Those delays are important in railroading. In passenger operation, if you miss the plane because your inbound connection was late, you have a delay of several hours; you get on the next plane and you go. The United States rail industry is often operating with 24-hour service headways; if the cars from E to B destined for this outbound train going to C miss that connection for whatever reason, this often causes a 24-hour delay until the next train—perhaps in some cases a 12-hour delay, but rarely less than that. Think about the impact of these delays on the total logistics costs of the affected receivers. Perhaps a stock-out results.

So, the operating policy to generate cost savings through consolidation implies stiff penalties when things go wrong. The reason that we run only one train a day from B to C is to achieve long train lengths. We pay a high fixed cost to run a train at all. Even if the railroad just operated the locomotive pulling one car, you would pay a fixed cost for the labor costs and for the locomotive utilization costs inherent in running even that very short train. (Fuel costs would be variable.) How do you achieve long train length? You do not run trains very often—perhaps once a day. You gather enough traffic so that you can operate long trains. This is fine from a cost viewpoint, but not from a service viewpoint. Your freight car misses the outbound train and you have a delay of a day. Also, if an outbound train is canceled because "there isn't enough traffic," you will always have delays. This is an example of Key Point 14, the cost/level-of-service trade-off.

Operating Costs

Refer to Figure 14.3. Train operating costs as a function of train length is shown. There is a high fixed cost and a modest variable cost—that is, the cost of running the train at all is high, but the incremental costs of adding a car are modest. We see a jump in operating costs when we need additional power units.

There are limits to train length because of safety and control issues, the length of sidings, and other operating questions. The relationship between cost per car and train length in cars is shown in Figure 14.4.

The cost per car on a long train is clearly going to be much lower than the cost per car on a short train. So there is an incentive to run longer trains from a cost point-of-view; that is what drives this idea of low train frequency on the United States railroad system. If one decided to run trains between B and C twice a day rather than once a day, that would improve service quality. If we run two trains a day between B and C with 50 cars on each rather than one train with 100 cars, there is a higher level-of-service associated with a higher train frequency; however, from a cost point of view it is more expensive to run two 50-car trains rather than one 100-car train.

FIGURE 14.3 *Train operating costs versus train length.*

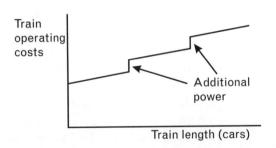

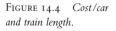

FIGURE 14.4   *Cost/car and train length.*

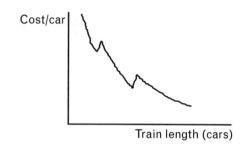

Operations Versus Marketing Perspectives

Now this simple idea connects to this tension between the operating and the marketing people. Who is going to want to run the 100-car train? The operating person or the marketing person? And who is going to want to run the two 50-car trains?

The operator wants to run long trains and the marketer the short trains.

Right—the operating person wants to run the 100-car train. The marketing person wants to run the two 50-car trains. The marketing person says, "I do not really care what this costs. This is better service for my customers and my bonus is tied to gross revenues; I am going to push for those 50-car trains; if I could convince them to run 100 'one-car trains', I would do that!" The operations person says, "I am measured on cost; I would love to run one train every two days with 200 cars. We would save money and what the customers want is a secondary consideration."

Now, obviously, that is stereotypical. In practice, the vice-president of marketing recognizes that his strategy would lead to higher costs and eventually higher freight rates, which he would not favor. The vice-president of operations would recognize that her strategy would lead to poorer service levels and eventually less traffic, which is not in her interest.

Train Dispatching

A related issue is train dispatching (see Figure 14.5). Suppose there is traffic coming in from A and E, destined for C.

FIGURE 14.5
Dispatching choices.

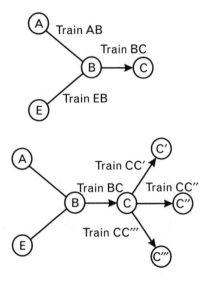

Train BC is scheduled to leave at 10:00 p.m. Suppose the train from A to B has arrived at C and has been processed so that traffic destined from B to C is on the outbound train BC. However, Train EB is late with its traffic from E destined to C. The dispatcher at this terminal has some choices. He could dispatch train BC now with, let us say, 60 cars on it—that came in from A through B destined to C—and leave the traffic from E to B behind to wait for the next day's train. Or he can wait 6 hours for the arrival and processing of Train EB. Then, instead of running a 60-car train, he will run, say, a 90-car train—60 cars and 30 cars, destined from A to C and E to C, respectively.

The Choice in Dispatching

Here is the choice. He can wait and run a longer train. He can dispatch now and run a shorter train. The first strategy gives him a more "economical" train. It is longer—it is 90 cars. So, on a cost-per-car basis, he is doing better. However, if he does wait for the late train from E to B, the 90-car train arrives at C late rather than on time as with the 60-car train; that could cause systems problems. For example, terminal C may require the power that was used on the train from B to C to dispatch some other train. Train BC is not here. It is 6 hours late and the power is 6 hours late. The train leaving C is going to be late as well as a result.

Further, there is more to the network; those cars destined for C′ and C″, came into terminal C late by 6 hours; the manager at terminal C will be faced with the same kind of question. That is, there was some other traffic that came from C''' destined for C' through C. The train coming in from B to C is 6 hours late because it waited for the traffic from E to B, and now this terminal manager is faced with the same question. Does she wait for train BC that is coming in late or does she dispatch CC' now, as a short train?

Holding for Traffic

So, the same issue that led to the terminal manager at B making a particular decision will impact the terminal manager at C. In fact, the reason that the original traffic going from E to B was late coming into node B was probably caused further west on the network. Somebody out west made a decision to delay a train—"holding for traffic," as the railroad people call it. The decision propagates down the network in a way that leads to instability in the way the network operates. So, holding a train is not a simple decision. One must consider network operating questions.

You recall in our simple elevator example, we talked about network behavior; in that case we had a rather simple network. But now we do not. We have a complicated network with a lot of connectivity and network behavior is critical. Judgments made at particular terminals can affect the operations throughout the network in ways that may be very difficult to predict.

How important is it that the cars may be delayed for a day?

Logistics theory helps us answer this. It depends on many different factors. The issue is that this late shipment could cause a stock-out at the umbrella store or at the General Motors assembly plant. The impact of that shipment being a day late depends on the kind of shipment it is; logistics theory is what helps us predict and understand what the importance of those delays is to customers.

Delay Propagation on Networks

Network behavior and how delays can propagate through the network is a general transportation issue—it is, of course, not limited to rail. The airline industry is very familiar with this. When you get bad weather in Denver, the propagation throughout the national system is substantial. Indeed, one of the motivations for building the new Denver airport, with its better capability for operating in the chronically bad weather conditions in that region, was the improvement in overall network behavior.

Network Stability

These networks can be quite unstable. Contrast air versus rail operations: a manager at the CSX railroad pointed out an idea that is, in some sense, obvious. Specifically, he noted that the airline industry has the same kinds of network instabilities as rail. If you measure network performance on passenger airline systems, it tends to be fine at 7:00 a.m. and deteriorates over the day. But, then 11:00 p.m. arrives—the system shuts down for 7 hours or so during the night. So you have a chance to rebalance the system. At the beginning of the next day you start with a system that is stable again.

In rail freight, you do not have that opportunity. The operation is 24 hours a day. So, the rail industry does not typically have that same opportunity to catch up until you get to the weekend, where the levels of freight go down substantially and the system has a chance to restabilize for the next week; the passenger air industry has that opportunity for 7 or 8 hours every evening. That is an interesting insight.

Back in high school physics, you learned about stable versus unstable equilibrium. Some of you will remember the ball at the bottom of the trough is in stable equilibrium. By this we mean if you take it out of equilibrium it tends to return by gravity to the same equilibrium point. The ball on top of the mountain is in unstable equilibrium; you push it out of equilibrium and it does not return to the same equilibrium point (see Figure 14.6).

How does one design systems that have stable rather than unstable equilibria? Design a rail network so when it moves away from equilibrium, the mechanisms that come into play return it to equilibrium.

FIGURE 14.6
Stable versus unstable equilibrium.

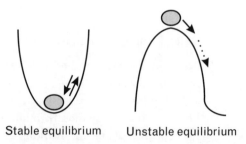

Stable equilibrium   Unstable equilibrium

In fact, it seems like the mechanisms that come into play tend to accentuate the disequilibrium and the network operates badly.

....

Operating Plan Integrity

Now, perhaps you are thinking at this point, "Why don't the railroads just run the trains on schedule as an optimal strategy?" In fact, this has been the recent management philosophy on some of the major U.S. railroads; it remains a point of great debate in the industry. The term in current use is "operating plan integrity" or sometimes "running to plan." The basic notion is design an operating plan that is feasible and makes sense. You have enough power; you have enough line capacity; you have enough terminal capacity to make this plan actually work. You run the trains according to plan, that is, according to schedule.

In reality, that is not the way it works. The schedule is not like a schedule for an airline or a schedule for a commuter railroad. Freight schedules in many cases are largely advisory. They are guidelines for terminal managers. The terminal manager is not necessarily expected to run according to the schedule. The terminal manager is supposed to make a decision about whether to hold a train at B for the EB traffic or not, utilizing the available terminal resources. You cannot be expected to be making those dispatching decisions while at the same time be expected to run on schedule.

Scheduled Versus Flexible Operation

So the big debate going on in the industry is whether the best strategy is running to plan in a disciplined manner. Some railroads feel that is an inflexible, uneconomic way of running the system. These railroads feel that flexibility for the terminal managers is useful and they can do a better job of balancing service and costs than they can by inflexibly running to plan.

Another approach tries to strike a balance between scheduled and flexible organizations. The railroad will periodically—perhaps daily—change the operating plan to reflect current conditions—weather, traffic, and so forth—and then run according to that modified plan [1]. Indeed, some railroads state that a disciplined operation is one that actually adheres to that modified plan.

Different railroads with different customers need to operate in different ways. Railroads that tend to have a higher fraction of high-value traffic with service-sensitive shippers are tending toward running to plan—a more disciplined operation. Railroads that carry less service-sensitive traffic tend toward more flexible operations so as to minimize costs. Indeed, some railroads will have different operating plans for different kinds of traffic.

So what does scheduled versus flexible mean? Consider Figure 14.7.

Daily Modified Operating Plan

Suppose we have developed, through optimization methods, an operating plan which governs the network. Suppose each day at 6 a.m., railroad management takes that operating plan and from it produces a daily modified operating plan which governs the way the railroad will operate on that particular day. The operating plan is a base case; the daily modified operating plan is a plan of action for a particular day. The daily modified operating plan takes account of stochastic conditions on the network like weather and traffic conditions. It also reflects how much a railroad is willing to change that base operating plan to accommodate conditions on a particular day.

There are several important issues inherent in this framework. If we assume the base operating plan was optimized, do we know how to change the base operating plan to produce an "optimal" daily modified operating plan? Can we do it fast enough—that is, in real-time—so that we can actually run our railroad? So real-time reoptimization is an important idea in this framework.

A second issue: if there is a big difference between the daily modified operating plan and the operating plan, can the terminal manager, given the resources that she has available, change her operations fast enough to operate close to optimally? Put another way, the instructions

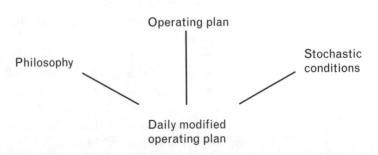

FIGURE 14.7   *A framework for transportation operations.*

come from central management that describe today's daily modified operating plan; this has implications for particular terminal managers who may simply not have the resources to operate according to that daily modified operating plan, particularly if it is radically different than the operating plan around which their terminal plan was designed in the first place.

How to Define Scheduled Versus Flexible Railroads?

Both of these are important questions, but it leads to another question—what is the definition of scheduled versus flexible for a railroad? There are (at least) two ways of thinking about this.

The first way: one could measure the difference between a scheduled and flexible railroad by the difference between the base operating plan and the daily modified operating plans that are developed for each day. A scheduled railroad would be one in which the operating plan and the daily modified operating plan were exactly the same. Our operating plan is relatively static; it may change only monthly or quarterly. In that way, the local terminal managers can plan, with relative certainty, what they will be asked to accomplish. However, at the same time, we are *not* accommodating to conditions like major changes in traffic, for example.

The second way: a scheduled railroad is one in which there is no difference between the daily modified operating plan decided upon at 6 a.m. and what they actually do that day. So, a scheduled railroad means that they adhere to the daily modified operating plan agreed to at 6 a.m. this morning, which may be rather different than the static operating plan.

This is a somewhat more modest definition of scheduled than the first one, but is one that, while accommodating to some stochasticity in the system, does adhere to a modicum of regularity by actually following through on the plan developed each morning. Some railroads do not even do that. The plan developed each morning is advisory and they may deviate substantially even from that.

REFERENCE

1. Dong, Y., "Modeling Rail Freight Operations under Different Operating Strategies," Ph.D. Thesis, Department of Civil and Environmental Engineering, MIT, September 1997.

CHAPTER 15

Railroad Terminals: P-MAKE Analysis to Predict Network Performance

Terminals

It is important to understand how terminals operate under conditions where trains come in on schedule, or perhaps not, depending on the decisions of the upstream terminal manager. Terminal performance is a major determinant of network performance. We want to ensure cars get processed correctly and go out on the right outbound trains in a timely fashion. Terminal performance is a function of the variability of incoming trains as compared with scheduled arrivals.

A terminal manager, if he or she is confident that schedules are going to be adhered to, can make some intelligent plans about how the terminal is going to operate. On the other hand, if trains are coming in at random—3 hours early or 5 hours late—it is much harder to plan for utilization of resources and operate effectively. So, we hypothesize a relationship as shown in Figure 15.1.

FIGURE 15.1
Terminal performance.

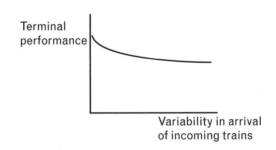

Terminal Robustness

The more variability there is in the arrival of incoming trains, the less effectively the terminal is going to operate. There will be some degradation in terminal performance. Cars will miss connections. More labor hours will be used.

We might design a terminal plan, including resources assigned to that terminal, which would be a very high-performance plan at low variability. If we could depend on the flow of inbound trains being timely and reliable, the terminal would operate beautifully. On the other hand, one could have a terminal plan that is a lower-performance plan but is more robust; the degradation of terminal performance as variability increases is more modest (see Figure 15.2).

The high-performance plan is "sensitive"; the low-performance plan for operating the terminal is "robust." An analogy is a Jaguar versus a Chevrolet. A Jaguar is a very high-performance automobile if perfectly tuned and if operated exactly the way it is supposed to be operated. The Chevrolet is not as high-performance a system. It does not go as fast, but it may degrade in performance under sub-optimal maintenance more gracefully than the Jaguar. So at low variability, when everything is perfect, the low performance Chevy does not look quite so good. But over a broad range of variability, maybe you do better with the more robust system—in this case the Chevrolet.

FIGURE 15.2
Terminal performance:
another look.

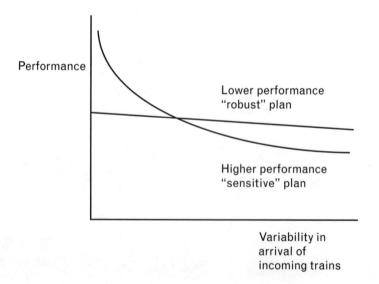

Performance includes a measure of cost—that is, if one is measuring terminal performance not only on throughput of the terminal but also on the resources used—robustness may involve a more conservative use of resources. It may involve having redundancy in the system, in order to assure that the system will operate even if the inbound trains have high variability. One can conceptualize terminal performance in this way; it is connected to the concept of "stable" versus "unstable" equilibrium—there is a linkage between how robust our terminal plans are and how stable the network is in its entirety. One would need robust terminals to have stable network performance.

Centralized Versus Decentralized Decision-Making

Another operating issue is centralized versus decentralized decision-making. For example, are train dispatching decisions made by individual terminal managers, or by a central controller for the network? All network-oriented transportation systems are, in practice, mixtures between central control and decentralized decision-making. In principle, central control is superior if we have perfect information available centrally in real-time. Those conditions do not usually apply; local managers on the site are often the only ones who can change operations in real-time in response to rapidly changing situations. As a practical matter, we need a mixture between decentralized and centralized decision-making for complex transportation networks.

LOS and Routing over the Rail Network

Level-of-service in rail freight operations is a function of the number of intermediate terminals at which a particular shipment is handled. One would expect that the LOS on a shipment that was handled at one intermediate terminal from origin to destination, would be better than the service for a shipment that was handled more than once, even if the total travel distance is the same. Indeed, empirical research shows the major determinant of the LOS is not the distance between origin and destination, but rather the numbers of times the shipment was handled at intermediate terminals, which is really an operating decision on the part of the railroads. One could provide direct service between Point A and Point C on the network, as shown in Figure 15.3.

By bypassing Terminal B, you provide direct service between Terminal A and C, which is likely to be faster and more reliable. The

Figure 15.3
Direct service.

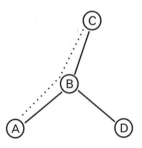

distance has not changed, of course, but given that the shipment was not processed at B, the level-of-service improves.

Terminal Operations

We continue by describing terminal operations, and then modeling the way in which networks operate as a function of terminal performance. Consider the following schematic of a classification yard (see Figure 15.4).

A Hump Yard

An inbound train arrives in the receiving yard. Many of our modern yards are "hump yards", that name derives from physically having a hump, or a gravity feed. An inbound train is taken over the hump; individual cars flow down into the classification bowl and are sorted—the term "bowl" is used simply because it is physically below the hump to allow the gravity feed to take place. This is simply a system for sorting out cars on inbound trains into new outbound trains for subsequent destinations. After the classification bowl, we have the departure yard, which again is simply a set of tracks onto which cars that have been

Figure 15.4
Classification yard.

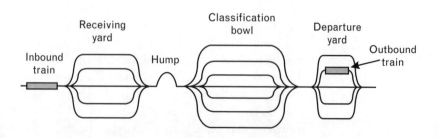

classified on various tracks are assembled into trains that will go out on the main line.

Queuing

Now, there are many issues in this kind of operation. As noted in the elevator example, queuing occurs in these facilities. For example, the hump is a resource; one train at a time can be humped, so trains may have to wait to be processed over the hump. Inbound and outbound inspections of these cars are performed by inspection crews; crews are resources. Often trains have to wait for an inspection crew. Resources are needed for assembling the trains—we have switch-engines in the yard that physically pull cars from the classification bowl into the departure yard and assemble them into trains. These are complicated operations. There are hundreds of cars and dozens of locomotives, and many crews in a major classification yard. There is queuing for resources; there is probabilistic behavior.

Micro-Simulation

In the rail industry there are many detailed computer micro-simulation packages of freight yards that help analyze what will happen if we change, for example, the pattern of inbound trains; or the outbound train schedule; or we improve the efficiency of the hump; or we add inspection crews. One can do sensitive, micro-simulation analyses to understand how this complex system behaves.

A Macro Perspective

Here we will take a more macro perspective in studying the impact of terminal operations because we are ultimately interested in the overall origin-destination service offered to customers.

Think About the Customer

Customers do not care about the operation of an individual terminal. What they care about is average transit time, reliability, loss and damage, service frequency—all those parameters we introduced earlier. So customers are interested in the performance of the network as a whole; here we will emphasize an understanding of how we can model the operations of terminals at a macro scale so we can predict the origin–destination performance of the network as a whole.

P-MAKE Analysis[1]

From a macroscopic perspective, a useful way of looking at a terminal is to consider the probability of an inbound car, say on Train AB, making its appropriate outbound connection on Train BC. Missing the connection has important service implications.

On a macroscopic level, the behavior of a terminal can be modeled by plotting a "P-MAKE" function, which represents the probability of a car on a particular inbound train making its connection as a function of the available time in the terminal until its planned departure on some outbound train. An example of a P-MAKE function is shown in Figure 15.5.

Available time to make that connection is defined as the time between the inbound train schedule and the outbound train schedule. So, if there are 8 hours to make that connection, the probability that that car will be on that outbound train is derived from the P-MAKE function, where the numerical value of P-MAKE is a function of AVAIL—the available yard time.

What are the reasons the car might not make that connection? This follows from the complexity, the queuing behavior, the probabilistic behavior, of the terminal facility. Let us assume that the P-MAKE function is as shown in the previous figure. Let us suppose if one has less than a minimal amount of time to make that connection—say, two hours—the car misses the connection. On the other hand, if one has some reasonable amount of time to make that connection—maybe 12 hours—one will make it with certainty. Between 2 and 12 hours, there is a relationship that describes the probability of making that connection as a function of AVAIL.

Now, given this relationship, we can derive how long, on average, a car is going to stay in the terminal. The amount of time that cars stay in the terminals is a substantial portion of the time it takes to get from origin to destination.

---

1   Carl Martland did the first work on this concept in his thesis, "Origin to Destination Unreliability in Rail Freight Transportation," Master of Science in Civil Engineering/Civil Engineer, MIT, Cambridge, MA, June 1972.

FIGURE 15.5   *A P-MAKE function.*

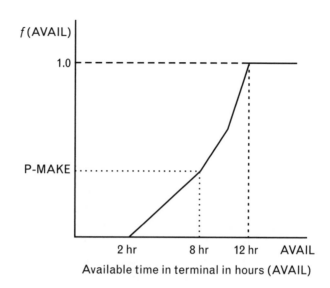

Available time in terminal in hours (AVAIL)

Average Yard Time

Let us compute the average yard time for a particular connection. Now, average yard time—E(YT)—will be a function of the available time (AVAIL) to make that connection. In this model, E(YT)—the average yard time—will have two components—the time spent in the yard if the connection is made, in which case, with probability P-MAKE, the terminal time is AVAIL. With probability (1 − P-MAKE), the car will spend (AVAIL + the time until the next possible train).

$$E(YT) = P\text{-}MAKE * (AVAIL)$$
$$+ (1 - P\text{-}MAKE) * (AVAIL + \text{time until next possible train})$$

Let us suppose that we have scheduled 10 hours for the car to make its connection from inbound to outbound train, and that for 10 hours, P-MAKE = .8—the probability that the car will make that connection and hence spend 10 hours in the yard. With probability (1 − P-MAKE) = .2—if we have daily service, a missed connection implies a 24-hour delay—so, with probability = .2, the car will spend 34 hours (10 + 24) in the yard. So, E(YT) = 14.8.

Now, suppose we plot the average yard time for a car as a function of the available time for connection between the inbound and outbound trains. As AVAIL gets bigger, we have a better probability of making that connection, but the car spends a long time in the terminal even if the connection is made. As AVAIL gets smaller, the yard time is small if the car makes the connection. However, it usually does not. Somewhere in the mid-range of AVAIL, we find a good place to operate which properly balances making and missing the connection (see Figure 15.6).

One can calibrate curves of this sort and come up with an optimal AVAIL for a particular terminal.

P-MAKE functions and, hence, appropriate AVAILs, will differ from terminal to terminal. P-MAKE functions may vary as a function of the terminal technology, as a function of the ratio of volume to capacity, and so on. Given this kind of information about the terminals in your network—you may have a terminal with the "sweet spot" between 10 and 12 hours, and you may have another with a "sweet spot" between 16 and 18 hours—one could, in principle, come up with a schedule that is built around the capabilities of those terminals.

What we are doing in this approach is modeling terminal performance without operating a micro-simulation. The P-MAKE function is an approximate representation of terminal behavior, which we can use for network analyses without performing a micro-simulation of every terminal every time we want to investigate a schedule change. It illustrates the modeling notion of trying to abstract complex portions of the system—in this case a terminal—into a tractable, mathematical

FIGURE 15.6
Proper balance.

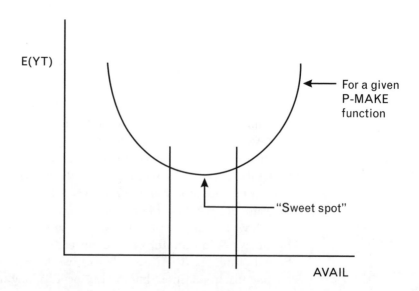

formulation that will allow us to analyze still more complex systems—in this case, networks.

P-MAKE Analysis

Now, we use the concept of P-MAKE to model a simple network to develop an understanding for how terminal performance, or yard performance (we use "yard" and "terminal" interchangeably) relates to the network level-of-service that the customer's shipment receives. Here is a simple network (see Figure 15.7).

Assume that the line haul times between each pair of terminals is 12 hours and that it is deterministic. Further, assume we have two identical intermediate yards—the P-MAKE function is the same for both these yards. Of course, there is no reason to expect two yards would have the same P-MAKE function. You might have a very modern, efficient terminal facility that processes cars very quickly. In that case, the P-MAKE function would be shifted to the left. Or, you might have an old terminal which has insufficient room and cars occasionally derail, in which case P-MAKE would be shifted to the right. We would tend, in this second yard, to make fewer connections for a given available yard time (see Figure 15.8).

But, for now—for simplicity—we assume these terminals are identical. Further, assume the trains have been scheduled so that there are 8 hours of available yard time in each terminal—8 hours of time for the connections to be made at each of these terminals for a car going between this origin-destination pair. Assume a one-train-a-day frequency; so there is a 24-hour headway between consecutive trains. If you make the connection, you will have 8 hours of yard time at each terminal. If you miss the connection, you have 32 hours of yard time, because you wait 24 hours for the next train, which we assume we make with certainty—probability = 1 (see Figure 15.9).

FIGURE 15.7 *Origin-destination performance.*

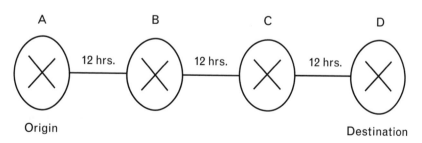

FIGURE 15.8
P-MAKE functions.

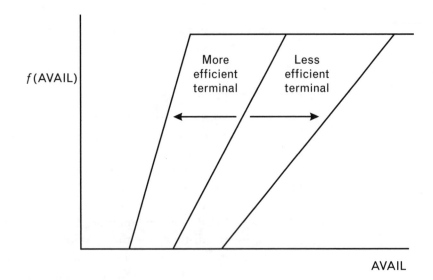

Origin-Destination Trip Times

We now compute origin-destination trip times. There are three possibilities for missed connections. First, the shipment misses no connections. It goes through both yards in 8 hours, making the outbound connection that it was expected to go on. Second, it could miss one

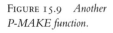

FIGURE 15.9 *Another*
P-MAKE function.

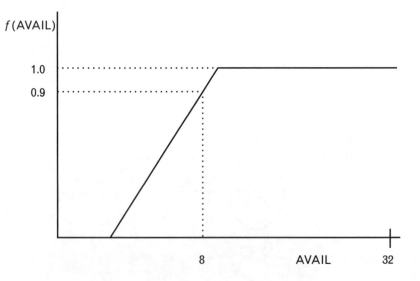

connection and make the other. Third, the worst case is missing both connections—it fails to make its connection at either yard.

Now, assume that the probabilities of making the connection at the two yards are independent. This is a questionable assumption. We introduced unstable equilibrium in transportation networks—we could imagine that the probability of missing the second connection, given the first one was missed, is possibly greater than it was if that first connection was made. But again, for simplicity we assume the probabilities are independent. The total yard time and the associated probabilities are as shown in Table 15.1.

So, for example, for $f(8)$ = P-MAKE = .9 (as in Figure 15.9) and, for a second case, $f(8)$ = P-MAKE = .8, the distribution of total yard time is shown in Figure 15.10.

To obtain O-D trip time, simply add 36 hours (3 links * 12 hours per link, which we assumed was deterministic) to total yard time. Note that the mean and variance of O-D trip time both rise as P-MAKE deteriorates from 0.9 to 0.8.

$$f(\text{AVAIL}) = \text{P-MAKE} = 0.9$$
$$\text{Average O-D Time} = 56.8$$
$$\text{Variance O-D Time} = 103.7$$

$$f(\text{AVAIL}) = \text{P-MAKE} = 0.8$$
$$\text{Average O-D Time} = 61.6$$
$$\text{Variance O-D Time} = 184.3$$

Are you saying available time is fixed for each terminal?

No, AVAIL is fixed for the *particular train connection* we are examining. Consider the following situation. Suppose Trains BC and EC are

TABLE 15.1   Yard Time Probabilities

Missed connection	Probability	Yard time (for AVAIL = 8 for both yards)
0	$[f(\text{AVAIL})]^2$	16
1	$2f(\text{AVAIL})*[1 - f(\text{AVAIL})]$	40
2	$[1 - f(\text{AVAIL})]^2$	64

FIGURE 15.10
Total yard time as an
f(AVAIL).

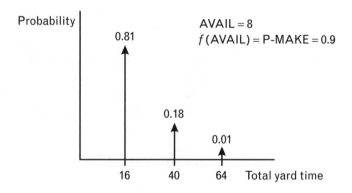

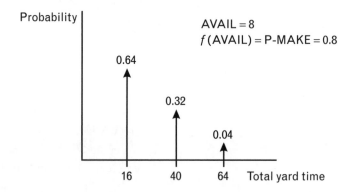

scheduled to arrive at 2 p.m. and noon, respectively, and Train CD is scheduled to depart at 10 p.m. Train BC has AVAIL = 8, relative to Train CD, while Train EC has AVAIL = 10, relative to Train CD (see Figure 15.11).

FIGURE 15.11
Available yard time.

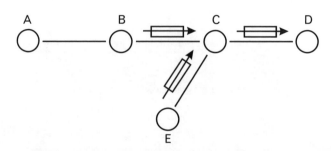

Train Frequency

Suppose we change train frequency to twice a day rather than once a day: 2 trains per day at 12-hour intervals. This changes the extra time in yards if we miss a connection from 24 to 12 hours. For example, in the case of AVAIL = 8 hours, with $f(8)$ = P-MAKE = .9, the yard time = 20 hours (8 + 12), rather than 32 hours (8 + 24). Again we assume that P-MAKE = 1.0 for AVAIL = 20 hours (see Figure 15.12).

The total yard time in this situation is shown in Figure 15.13. So, by having trains run twice a day, the average yard time and variance of yard time goes down. Note that we have not changed the efficiency of the terminal and how quickly cars can get through it. We simply changed our operating policy to 2 trains per day. Twice-a-day train frequency has cost implications. The railroad needs more power to run that system than it would to run a once-a-day system. The railroad needs more crews to run the greater number of shorter trains. This system is a more expensive system, but provides a better level-of-service. This is the classic cost/LOS trade-off (Key Point 14).

Bypassing Yards

Another option is bypassing a yard. Rather than the trains being switched in both intermediate yards, we could design the service such that the car is processed at the first yard—laying ourselves open to the possibility of missing a connection—and bypasses the second yard (see Figure 15.14).

Think of it as nonstop service. In an airline system, for example, you travel directly from Chicago to Boston, rather than having to change planes in Pittsburgh. Generally, you will have a faster and more reliable trip. You do not subject yourself to the possibility of the missed connection in Pittsburgh. But, from the airlines' viewpoint, the load

FIGURE 15.12
P-MAKE function with more frequent trains.

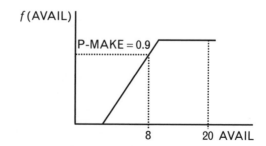

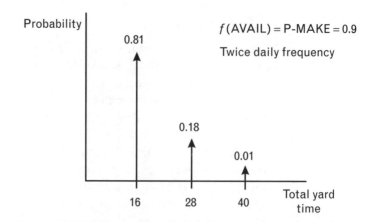

FIGURE 15.13 *Total yard time with more frequent trains.*

Average O-D time = average yard time + 36 = 52.96

Variance O-D time = 25.91

factor on the Pittsburgh–Boston link may be lower, because St. Louis–Boston passengers routed through Pittsburgh are not consolidated with Chicago–Boston passengers.

We use our simple model to study the yard bypass strategy; suppose the shipment goes through only one yard. If, for example, with $f(\text{AVAIL}) = \text{P-MAKE} = .9$, and once-a-day service, we have a probability of spending 8 hours in the yard with a probability $= .9$; with

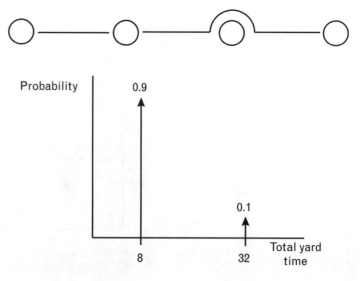

FIGURE 15.14 *Total yard time with bypassing one yard.*

probability = .1, we spend 32 hours. So again, the distribution of yard time is pushed to the left, improving average total yard time and variance of total yard time.

Next, one could bypass both yards—therefore totally eliminating the variability that is introduced by processing at the terminals. Under those circumstances, the origin-destination trip time has zero variance, given our assumption of deterministic link times. So, we have improved the mean value to 36 hours and reduced the variance to zero.

We have through service and shorter trains (because we still have to provide service from A to B; A to C; and so forth). But, by providing service without intermediate yarding between A and D—at the extremes of our network—we have a better level-of-service, but costs are higher. That trade-off is central to transportation operations planning regardless of mode.

CHAPTER 16

Car Costs and Level-of-Service

A More Subtle View of Costs

Let's now take a more careful look at this yard-bypass decision and ask whether it is really more expensive. Now we have to think more subtly about what the cost structure is. To simplify, consider only how much it costs the railroad to operate the service between the two extreme points of the network, A and D (we ignore the network effects), as shown in Figure 16.1.

The railroad is concerned with what A to D service costs to provide as a function of various operating options: skipping one yard; skipping two yards; once-a-day train frequency; twice-a-day train frequency; or, changing the performance of the yards by investing in better technology.

We subdivide operating costs into three categories: train costs; terminal operating costs; and car costs associated with the costs of owning rolling stock.

$$costs = train\ costs + terminal\ operating\ costs + car\ costs$$

Here we need not include the infrastructure costs. While there are costs associated with building and maintaining it, we assume that the

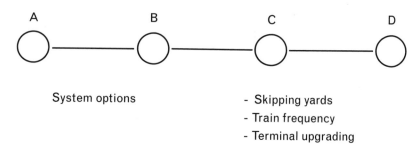

FIGURE 16.1 *System options.*

System options

- Skipping yards
- Train frequency
- Terminal upgrading

infrastructure costs, in the short run, are not a function of the different operating options. We assume that the costs to maintain the track and so forth will not change as we compare two versus one train a day.

Now, how might these cost functions look? If we look at train costs as a function of train length, the railroad industry likes to run long trains because you pay a high fixed cost to run a train of any length. So, we have a high fixed cost and a relatively small variable cost per car.

How about terminal operating costs? That has a similar kind of behavior; that is, terminal operations have a high fixed cost. Then, there are variable costs for operating the terminal that are a function of the number of trains to be processed and cars handled, as shown in Figure 16.2.

Car Costs

Now, we introduce the idea of "car costs." Freight cars cost money. To compute car costs for this A to D service, we introduce the concept of a car cycle. Key Point 4 suggests the car cycle is fundamental to transportation systems analysis. The vehicle—in this case, the freight car—is an important financial asset. It is the mechanism for providing service.

Freight Car Cycle

A freight car moves through the railroad network as follows:

• Railroad places car at shipper siding;

• Shipper loads car;

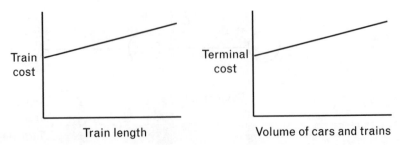

FIGURE 16.2   *Train costs and terminal costs.*

• Railroad moves car through system to destination;

• Receiver unloads car;

• Railroad picks up empty car and places car at next shipper's siding.

In equation form, we write:

car cycle (CC) = shipper's loading time + transit time (loaded)
+ receiver's unloading time + transit time (empty)

The loaded and empty transit time include time spent in terminals. We assume this car-cycle time is deterministic. Further, assume that we simply send the empty cars directly back from D to A.

Fleet Size Calculation

Assume we have a volume (V) of carloads per day going from A to D; we can compute the number of cars (NC) we need in the fleet to perform this service.

$$NC = V * CC$$

Now obviously, this is a substantial simplification. The system is not deterministic, but rather probabilistic. Second, empty cars often are sent to some point other than the one from which they came, as shown in Figure 16.3.

In a situation of car surplus—light traffic, with excess car capacity—freight cars may queue for loads.

Car Cost/Day Calculation

But, for illustrative purposes, we turn to our simple example. How would we compute the car costs associated with this fleet? There are many ways of thinking about it. We will start with a simple construct and then go to a more complex interpretation.

One could compute—given one knows how much was paid for the car and how many years the car is expected to stay in service—the ownership costs per day for that particular car. Multiplying it by

FIGURE 16.3   *Loaded and empty moves.*

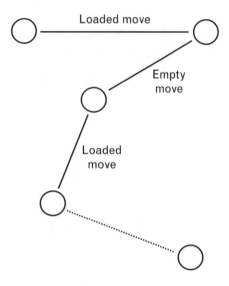

the number of cars you need to operate the fleet, one could come up with an ownership cost for your fleet as a function of the purchase price of the fleet.

Now, as cars get more expensive, the cost to the railroad of longer car cycles (CC) grows—because you have to provide a more expensive asset—that is, cars—to provide your customers with service. If one has a $15,000 freight car, this is a very different situation from an $80,000 freight car. Therefore, the kinds of decisions you might make about how to utilize that resource may very well be different.

The way you operate the system depends upon how you think about the costs of cars. Suppose the car cycle gets longer, as it would if we use intermediate terminals—for the A to D move—rather than direct A to D service. Earlier, we characterized direct service as expensive because the train is shorter. But when we have that direct service, we also have a shorter car cycle and, in fact, a smaller number of cars in our fleet for that service. Therefore, the car costs will go down:

$$car\ costs/day = NC * daily\ cost\ of\ owning\ a\ car$$

And the question will ultimately be—does the lowering of car costs offset the cost of providing this more direct service with shorter trains? Now, this is a subtle issue. For example, we can think about this from a long-term or a short-term cost perspective. If one looks at it from an economic perspective, one basically could say, "Look, it cost me $15,000 to buy that car; it will last so many days, and therefore my cost of ownership per day is $15,000 divided by that number of days."

On the contrary, someone else could say, "Look, that does not really make a lot of sense. You bought the car, and it is sitting around doing nothing; you have already bought and paid for it, so it does not really cost you anything at all." The car ownership costs can be thought of as sunk costs rather than costs that accrue on a day-by-day basis.

Cost Versus LOS

We plot an idealized relationship between cost and level-of-service (see Figure 16.4).

You provide a better service by running shorter, more frequent trains, bypassing yards, and so forth. The train costs go up with level-of-service because you have more and shorter trains. On the other hand, car costs will go down as level-of-service is improved. Why? Because the fleet is more productive when we bypass yards and run more frequent trains. Therefore, the fleet size needed to operate will be lower. So car costs will go down as you improve level-of-service. If you plot the sum of train and car costs, in principle, we can find an optimum level-of-service at which to operate, LOS*.

LOS* is the level-of-service that optimizes costs—without even considering that better level-of-service may attract more traffic.

FIGURE 16.4   *Cost versus LOS.*

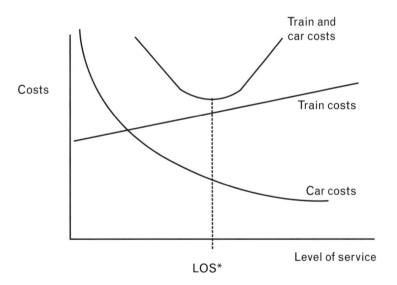

Should you not include terminal operating costs in that equation?

Good question. If we plot the terminal costs as a function of level-of-service, what do you think it would look like?

Terminal costs would go down, because we process fewer cars at better levels-of-service.

Right. We skip terminals. On the other hand you have more train dispatching overall. Terminal costs would tend to go down as the level-of-service improved as we bypassed yards. If we include terminal costs, the analysis does not change conceptually. We still have an LOS* that is optimal, from a cost point of view.

Now, let us suppose we have this relationship plotted. Consider how it would change if we retire the old fleet and buy a new one with cars that cost 50% more. What do you think would happen in this analysis? What would happen to LOS*?

It would move to the right.

Correct. The car cost curve would move up (see Figure 16.5).

The cost for owning a car for a day went up; it is more valuable to the railroad to save car-days by shortening the car-cycle, by providing a better LOS. It is cheaper, in this instance, to provide better service!

FIGURE 16.5 *Costs versus LOS—another view.*

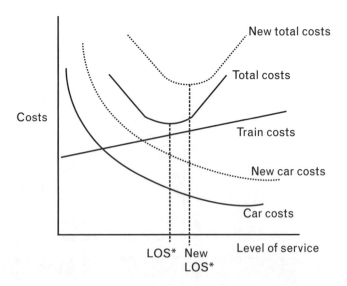

Now, suppose you meet your vice president of labor relations. She says, "Good news! I just finished negotiating a deal with our unions; instead of four-person train crews, I have just convinced them that we need only two people in a crew." What happens?

Fixed train costs go down.

Right. If train costs get cheaper, the optimal level-of-service moves to the right. It is cheaper to run more frequent trains, which also produces a better LOS.

A problem in the rail industry is that adjustments to these cost changes are slow. We have seen a substantial rise in the cost of rolling stock—freight cars—over the last decade. As the trucking industry has become more competitive, the shippers have pressured the rail industry saying, "You must provide us better cars; you have to buy new equipment, if you are to retain our business." So, car costs have risen; however, re-optimization of LOS* is slow to happen.

As car costs go up and train costs go down—both of which have occurred—you may in fact be operating at the wrong LOS from purely a cost viewpoint. You may very well want to operate, given current cost trends, at a higher level-of-service, even without consideration of the market benefits of doing so.

Another View of Car Costs

So far, we have simply taken the ownership costs of the car and divided it by a projected life to arrive at a daily cost of owning the car. One could argue that this cost is too low. Suppose we are operating during a strong economic period and that the railroad had an infinite queue of loads that needed service; the railroad can just keep servicing loads as long as it can provide cars for them. One could argue that car costs in this situation include not only the ownership costs but also the forgone profits when you do not have enough cars available to carry loads.

Contribution

Consider the concept of "contribution," which is the revenues that could be generated, say, per day, by that car, minus the costs of carrying that load per day; this gives a measure of the profitability of that particular move. Profits are forgone by not having cars available. We charge ourselves for cars on that basis, pushing car costs up further; there is an incentive to run a still shorter car-cycle, which again means a better level-of-service for customers.

How Performance Measures Affect Decisions

The way in which managers are charged for car use should affect their railroad operating decisions. And if they are not charged anything for car use, they may choose to run very long trains because it does not cost them anything—as their performance is measured—to delay cars a long time in the yard. On the other hand, if they are charged substantially for car use, they may run short trains and provide a better level-of-service at the same time. The way in which people are measured and the incentives they are provided will alter the way they operate (see Key Point 28).

Variability in the Car Cycle

We now extend our model to include unreliable (i.e., nondeterministic) loaded and empty moves using a simple two-node case (see Figure 16.6).

FIGURE 16.6
Stochastic car cycle.

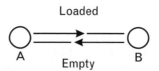

Loaded

A   Empty   B

"Unreliable" loaded and empty moves

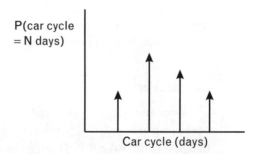

P(car cycle
= N days)

Car cycle (days)

Unreliability in Loaded and Empty Moves

This unreliability could come about in any number of ways. It could happen because there are intermediate yards between A and B. Or, perhaps the receiver might unload cars in a stochastic manner. In any case, there is unreliability—a variation. The car cycle is now not a deterministic constant; rather the car cycle is a random variable, as shown above. On average, it takes N days, but sometimes it takes N minus 1, and sometimes it takes N plus 1. What does that imply about the number of cars needed for the fleet to provide service? What kind of problem is that?

It is an inventory problem.

Car Inventory

Correct, it is an inventory problem, just like the inventory problem in our umbrella store. The railroad operator has an inventory problem. There are unreliable deliveries of empty cars back to the originating point. So, under the assumption that the railroad wants to provide all the cars that the shipper wants—say, 50 cars a day, every day, you need a bigger fleet. The railroad needs to buffer itself against its own reliability. So the size of the fleet needed to provide a continuous flow of cars to the shipper will go up under conditions of unreliability on the network—as opposed to the deterministic transit time on the network assumed earlier.

So unreliable operations cost the railroad money in several ways. It costs money because the railroad has to provide an additional inventory of empty cars. The loss is in addition to the lost business, because service was unreliable in the first place. The railroad makes a choice about the size of the inventory of empty cars, but it affects level-of-service in terms of availability of empty cars to the shipper—just as transit time on the networks affects level-of-service.

Railroad operating people think in very different ways about priorities for empty versus loaded cars on the system. They give priority to loaded cars if, for example, there is some limit on train length because of power constraints. After all, somebody is paying to move the car from A to B. However, it is not always necessarily the optimal strategy. Moving those empties to where they can be loaded is also important and unreliability in placement of empties matters. The rail industry recognizes more and more that one has to be concerned with the reliable redistribution of empty cars on the network.

CHAPTER 17

The Kwon Model—Power, Freight Car Fleet Size, and Service Priorities: A Simulation Application

Power, Freight Car Fleet Size, and Service Priorities

Oh Kyoung Kwon developed some ideas on service priorities in the rail industry [1]. Consider a two-node network. Shipments go from A to B with a two-day deterministic trip time; there is a one-day deterministic time at terminal B for unloading the cars, and then a two-day deterministic empty return (see Figure 17.1).

A shipper is generating loads into the system at node A; this shipper is permitted to assign priorities to traffic. The shipper has three kinds of traffic. Some of it is very service-sensitive; some of it is only moderately service-sensitive; and some of it is not service-sensitive at all. Service-sensitivity refers to requirements for mean trip time and trip time reliability.

FIGURE 17.1   *A simple network.*

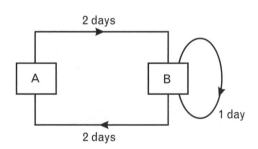

2 days

A

B

1 day

2 days

So, the shipper can designate traffic as high-, medium- or low-priority. The railroad will charge that shipper more for high-priority than for medium-priority and will charge more for medium-priority than for low-priority.

Let us further assume that the volumes of traffic at each priority level are probabilistic.

Power Selection

Suppose that the railroad has to make a decision about the locomotive power it will assign to this service, which, in turn, defines the allowable train length. So the railroad makes a decision, once-per-time interval—for example, a month—on the power that will be assigned to this service: power enough to carry 100 cars per day, 120 cars per day, 140 cars per day, and so forth. This is a capacity decision. A probability density function describes the total traffic generated per day in all priority classes: high, medium, and low. This would be obtained by convolving the probability density functions of the high-, medium-, and low-priority traffic generated by that shipper, assuming independence of these volumes. The issue is capacity selection, just as in the elevator system. We could decide that the design train length for this particular system is as shown in Figure 17.2.

FIGURE 17.2
Probability density function for daily traffic.

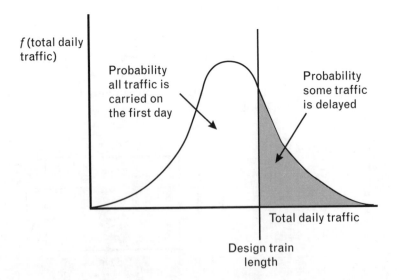

INTRODUCTION TO TRANSPORTATION SYSTEMS

On the first day, the probability of carrying all the traffic is equal to the area under the curve to the left of the design train length. That capacity line could be moved to the right by providing more power; a better level-of-service is provided because the railroad carries traffic on the day customers wanted it to be carried.

Car Fleet Sizing

In addition to power, the question of car fleet sizing affects capacity. There is an inventory of empty cars. While, in this particular case transit times are deterministic, we do have stochasticity in demand. So the railroad needs a different number of cars each day. The implication is that there may not always be enough cars if that fleet is sized without any slack.

So, there are two capacity decisions: power and the inventory of cars. In this particular case, the shipper has assigned priorities for traffic that we will use to make decisions about how to use that scarce capacity. We have pulling power on the locomotive such that some day we have to leave some cars behind; or we may stock out of empty cars at terminal A and we have to decide which loads are loaded into the cars we do have. Presumably, the shipper-assigned priority is a way of making those decisions.

Train Makeup Rules

Kwon discusses the idea of train makeup rules and looks at three of them.

Makeup Rule 1

The first train makeup rule was quite simple. You load all the high-priority traffic; then you load all the medium-priority traffic; then you load all the low-priority traffic; finally you dispatch the train. For example, suppose the power constraint is 100 cars, and that we have an infinite supply of empty cars.

Day 1 traffic
  High = 60
  Medium = 50
  Low = 60

Train
  High = 60
  Medium = 40

Traffic left behind
  High = 0
  Medium = 10
  Low = 60

Day 2 traffic
  High = 40
  Medium = 50
  Low = 50

On Day 2, using our first rule, we load the 40 high-priority cars first (even though they are Day 2 traffic). Then we load the 10 medium-priority cars from Day 1 and the 50 medium-priority cars from Day 2. We now have 110 low-priority cars left over, 60 of which have been delayed 2 days and 50 of which have been delayed 1 day.

Makeup Rule 2

Consider a second train makeup rule. First, clear all earlier traffic regardless of priority. Using the same traffic generation, the first day's train would be the same. On the second day, you would first take the 10 medium-priority and 60 low-priority cars left over from Day 1, and fill out the train with 30 high-priority cars from Day 2, leaving behind 10 highs, 50 mediums, 50 lows.

Makeup Rule 3

A third option strikes a balance, since we do not want to leave high-priority cars behind. In this option we:

  • Never delay high-priority cars, if we have capacity;

  • Delay medium-priority cars for only 1 day, if we have capacity;

• Delay low-priority cars up to 2 days.

In this case on Day 2, we first take the 40 high-priority cars, the 10 medium-priority cars left over from Day 1, and the 50 medium-priority cars from Day 2.

So, at the end of Day 2, we have:

• No high-priority cars;

• No medium-priority cars;

• 60 low-priority cars with 2 days of delay;

• 50 low-priority cars with 1 day of delay.

Suppose the Day 3 traffic is as follows:

High = 30
Medium = 60
Low = 30

So the train on Day 3 would carry the 30 high-priority cars, the 60 low-priority cars with 2 days of delay, and 10 of the medium-priority cars generated on Day 3 (since the 50 low-priority cars have been delayed only 1 day).

Comparing the Rules

Compare these various rules. Under the second train makeup rule, the railroad carries yesterday's low-priority traffic instead of today's high-priority traffic. Does it make sense? To understand that, you would need to go back to the total logistics costs model to see what those delays to the high-priority traffic are going to cost your shipper.

Kwon sought a balance. In his third strategy, when medium-priority has waited for 1 day, it, in effect, becomes high-priority. When low-priority has waited for 2 days, it, in effect, becomes high-priority. Of course, Kwon could have as easily said medium-priority traffic waits 2 days and low-priority waits 4 days. It is simply a way of balancing the capacity used for the various kinds of traffic.

Remember that there are different prices being paid for this service; the question of how you manage your railroad depends on the service requirements for those three classes of traffic and those requirements are related to price.

Service for Traffic of Different Priorities

This system generates service distributions that might look as follows (see Figure 17.3).

FIGURE 17.3 *Service versus priority.*

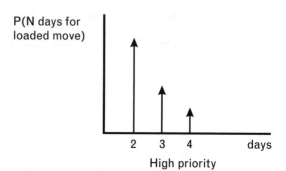

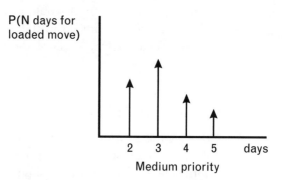

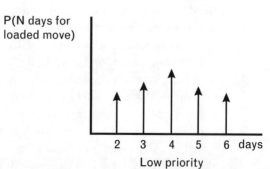

The levels-of-service will differ substantially because of the strategies we are using to allocate capacity. Now, what can you do to improve the service distributions? Buy more locomotives and increase train capacity. Or, buy more empty cars for inventory so that the probability of running out of empties is lower.

Do You Want to Improve Service?

But the question for the railroad operator is: for whom is service being improved? If you add that capacity by spending money on extra locomotive capacity or extra cars, there may be some improvement for high-priority traffic which gets excellent service anyway. There may be reasonable improvement for medium-priority traffic and a lot of improvement for low-priority traffic, because the right-hand tail of that distribution is eliminated and the mean value and variance drops.

Is this a good strategy? Maybe the low-priority traffic is really indifferent to better service. If the railroad plans to charge more for that service—which it presumably does, because it spent money to add capacity—it may lose the low-priority traffic which is price-sensitive and not service-sensitive. So when you think about adding capacity, you need to think about what service is being improved and whether, in fact, shippers will be willing to pay the price for that better service. The low-priority shipper may be perfectly happy with paying little for the mediocre service.

Allocating Capacity

We are allocating capacity on this system. This is the same as the earlier elevator example, where we discussed charging a congestion fee. If you wanted to use the elevator at a particular time of day, say at peak hour, you had to pay a premium. If you wanted to use it off-peak, you were charged less.

The same idea applies here. The railroad is, in effect, allocating capacity by limiting which traffic goes on that train. It decides on the basis of how much one pays for the service. And the railroad is pushing low-priority traffic off the peak, and in effect paying the low-priority customer the difference between what high-priority and low-priority service costs to be moved off the peak.

A Nonequilibrium Analysis

Earlier, we introduced equilibration as an important concept in transportation. We discussed supply functions, emphasizing the fact that the more traffic there was, the poorer the level-of-service was, given fixed capacity. We discussed demand, emphasizing that the better the level-of-service was, the more volume was demanded.

Understand, though, that the analysis we just performed is a nonequilibrium analysis. We assumed that the shipper just sits there without reacting. And, in fact, that is not the way the world works, if one applies microeconomics principles.

An example: Let us suppose the railroad was using the second of the train makeup rules. You clear out the Day 1 traffic before you take care of the Day 2 traffic. That mutes the difference between high, medium, and low service. The customer may note, "Medium-priority and high-priority service look pretty similar when I look at the histogram for transit times. I think I will move some of my high-priority traffic down to medium-priority because I do not want to pay a premium for high-priority service that is almost exactly the same as medium-priority service." Or arguing the other way, if we are using the first makeup rule, which says high-priority always goes first, you are getting a sharp distinction between services. Then the customer may say, "That high-priority service is really super. Transit time of two days. Day in and day out. They never miss. Paying the extra money is well worth it for this cargo for which I am currently using medium-priority service. So, I will move it up to high-priority because I will pay a modest premium for terrific service."

However, what happens when he does that? He changes the way the system operates. The railroad's plan is based on probability density functions for high-, medium-, and low-priority traffic. The railroad official concerned with high-priority traffic says, "High-priority traffic is getting strong. It was a mean value of 50 loads per day. Lately it is 75 loads per day. I wonder what is going on." His colleague says, "I have been noticing that medium traffic went from 50 down to 25. I wonder what is going on. Is our medium service poor? People are not using it." Maybe some day they have a cup of coffee together and they realize what is going on. The system is re-equilibrating, because high-priority service is underpriced. However, the levels-of-service will change when you have more high-priority traffic. The customer said, "Let us pay that small premium for better service." All of a sudden, the service is not quite as good anymore because more people are using high-priority service.

So, this simple analysis is handy to get a sense of how the system works, but understand it is a nonequilibrium analysis. We have not considered how shippers are going to react to pricing and service changes and how the railroad is going to react when shippers change behavior.

Many nonequilibrium analyses are quite legitimate analyses. In fact, very often as a practical matter, analyses of the sort that we did here turn out to work well especially in the short run, where the lack of equilibrium in the model causes no prediction problem.

Remember:

All models are wrong; however, some are useful.

Kwon's model is "wrong" but it is useful also. We do not need to have perfection on every dimension of our model in order to be able to use it effectively in an analysis.

Earlier, we abstracted a complicated terminal with cars and people and locomotives into a P-MAKE function. That is obviously "wrong." P-MAKE is a very simple function representing a very complicated terminal. However, it may still be very useful for understanding the system. So, we do not have to reach perfection with our models being extraordinarily detailed and precise. We only want models that help us in making good decisions about our system.

Investment Strategies: Closed System Assumption

How would you decide about investment strategies in this circumstance? How would you decide whether you want to invest more in locomotives or more in freight cars? One way to think about it is the closed system formulation. By "closed system" in this particular sense, we treat the shipper and the railroad as one company. It is a closed system in the sense that the price of transportation—what would normally be the rate charged by the transportation company to the shipper—is internal to the system. It is a transfer cost and does not matter from the point of view of overall analysis.

Then, if we choose the locomotive size and choose the number of cars in the fleet, we can compute operating costs. Also, we can—using the service levels generated by the transportation system operation with those locomotive and car resources—compute the logistics costs for the shipper. We use inventory theory and, if we make assumptions about the value of a commodity and the needs of the shipper from an inventory viewpoint, we can estimate the total logistics costs (TLC) associated with a particular transportation level-of-service. So, compute the operating costs and compute the logistics costs—absent the transportation rate, which is a transfer cost in this formulation—at a particular resource level and at a particular level of demand for high-, medium-, and low-priority traffic.

Then you optimize. Change the capacity of the locomotive; change the size of the freight car fleet; and search for the optimal sum of operating costs plus logistics costs. Under the closed system assumption—that transportation rates are simply an internal transfer—you could come up with an optimal number of cars and an optimal number of locomotives for the given logistics situation.

Allocating Costs to Priority Classes

If you relax the assumption about the closed system—now there is a shipper that is paying money to the railroad to carry his freight, rather than an integrated firm—the railroad is faced with another difficult problem. If that railroad wants to make money at all priority levels—high, medium, and low—the railroad has to be able to compute how much it costs to provide high-, medium-, and low-priority service.

The allocation of costs to high-priority, medium-priority, and low-priority traffic is nontrivial. For example, suppose you have a large inventory of cars to ensure that you almost never stock out. This means some cars in your fleet are rarely used. You still have to pay the price for owning those cars. How do you allocate the cost of those cars to the high-, medium-, and low-priority services? This is a difficult question, as discussed in Key Point 13.

Simulation Modeling

How did Kwon actually compute his operating and logistics costs? This formulation is a hard probability problem to solve in closed form. What you would like to be able to do is use this probabilistic formulation of a railroad system and compute a probability density function that described the sum of operating costs plus logistics costs (see Figure 17.4).

The cost for 30 days is a random variable. There is some probability density function that describes the operating costs and logistics cost of that system for a 30-day period. So, what we would like to do is compute this probability density function in closed form from the stochastic inputs.

But, we do not know how to solve this problem in closed form. We need to use the technique of probabilistic simulation to generate results. Probabilistic simulation is based on the concept that through a technique called random number generation, one can produce variables on a computer that we call pseudo-random. They are not truly random since we are producing them with a computer program, typically by multiplying together two very large prime numbers and using the remainder as a random variable uniformly distributed on the interval [0,1] (see Figure 17.5). Through this device, we obtain streams of pseudo-random numbers that allow us to simulate random behavior. We map those pseudo-random numbers into random events.

If we have an event that occurs with a probability equal to .4, we generate a pseudo-random number on the [0,1] interval. If that random

FIGURE 17.4
Probability density function of 30-day costs.

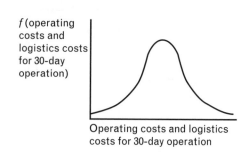

FIGURE 17.5
U[0,1] distribution.

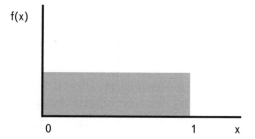

number is less than 0.4, the event happens. If it is greater than 0.4, the event does not happen.

The numbers are pseudo-random in the sense that if the random number generator is reinitialized, the same stream of numbers occurs. If well designed, the numbers have little serial correlation so one can use them as though they were truly random, independent from one another and uniform on the [0,1] interval.

We can write computer programs that generate 30 days' worth of traffic using a random number generator and run the trains, and then compute the logistics costs and the operating costs for that 30-day period.

At the end of the 30 days of simulation, you have one sample from the probability density function for operating costs plus logistics costs, which is a random variable. In order to trace out that probability density function, one would have to run 30 days, zero out the system and run 30 more days, and so on, until we could compute a mean value of operating costs plus logistics costs with some statistical confidence (see Figure 17.6).

Simulation Versus Probabilistic Analysis

Simulation is a technique that is very useful in transportation systems analysis, since it helps solve probability problems that cannot be solved

FIGURE 17.6
Simulation as sampling.

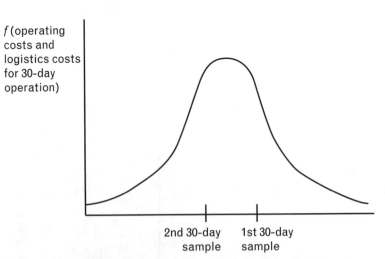

f (operating costs and logistics costs for 30-day operation)

2nd 30-day sample

1st 30-day sample

Operating costs and logistics costs for 30-day operation

in closed form. And given the complexity of probabilistic formulations for many transportation systems, this tool is very valuable.

Figure 17.7 illustrates the distinction between the analytic approach where the model is solved, and a simulation approach where the model is operated.

Probabilistic analysis is a good tool, but to actually develop useful results, we often have to simulate, which can use a lot of computer time. You may have to do a lot of repetitions to get statistical confidence. There is uncertainty in the results. Of course, you would much rather have a closed-form solution. You would rather solve for the probability density function that describes the random variable of operating costs plus logistics costs—only you cannot without making gross modeling assumptions. So, what you do is simulate using a lot of computer time, because it is often the only viable option you have.

REFERENCES

1. Kwon, O. K., *Managing Heterogeneous Traffic on Rail Freight Networks Incorporating the Logistics Needs of Market Segments*, Ph.D. Thesis, Department of Civil and Environmental Engineering, MIT, August 1994.

2. Law, A. M. and W. D. Kelton, *Simulation Modeling and Analysis*, 2nd ed., New York: McGraw-Hill, 1991.

FIGURE 17.7
*Analytic versus
simulation approach.*

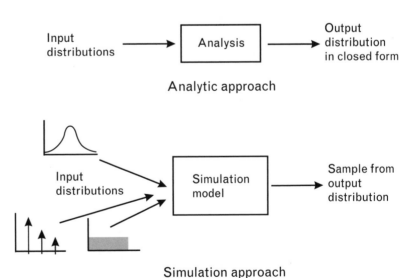

Analytic approach

Simulation approach

CHAPTER 18

Measuring Origin-Destination Service and Other Rail Issues

Measuring Origin-Destination Service

How is origin–destination service measured? Consider a hypothetical histogram for a particular origin–destination pair on a railroad (see Figure 18.1). This plots the probability that the trip from origin to destination will take some number of days, N.

The classic statistical method of characterizing this distribution is to compute mean and variance. However, with real-world data systems, variances are often not especially useful, because the distribution will have points on the right-hand tail that will have a large effect on the computation of variance, but that often turn out to be data errors. These

FIGURE 18.1 *Trip time histogram.*

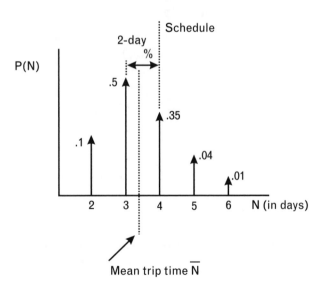

extraordinarily long trip times, if used without some screening, will lead to an overestimate of variance, because the very long trip never happened. So variance, while the classic measure, is often not the most useful one.

Schedule Adherence

A measure that is often used is schedule adherence. If the schedule called for a 4-day transit time between origin and destination, we measure the number of cars that came on time or before. Schedule adherence has several problems. First, it does not address early arrivals. Early arrivals may be considered negative by the receiver of the goods. The shipment arrives and the receiver has no place to store it, for example.

The other problem with schedule adherence as a measure is that it can be easily manipulated. One can relax the schedule and apparently perform better in the sense that a larger number of cars are now on schedule. But nothing has really changed.

This is a chronic problem in the airline passenger industry. When the Federal Aviation Administration publishes data about schedule adherence, a not-unreasonable response on the part of the airlines is to change their schedules—making scheduled trip times longer. The operation may not have changed at all. But performance looks better on a schedule adherence basis because the schedule is easier to adhere to.

Very Bad Trips

Often, shippers are concerned about very bad trips that are not data errors. So, for example, consider moves that take five days or more, as a variability measure for this distribution. We want to measure the right-hand tail. We do not weigh these moves by their distance to the mean; we only say that 7% of the moves took more than five days. The underlying idea is that stock-outs are often caused by those very bad moves. Again, this measure can be manipulated by changing the cut-off point, by using six days instead of five, for example. You do not change the performance, but you change the apparent performance.

"X"-Day Percent

A measure that is useful in the railroad industry is an "X"-day percent. Suppose we are using a 2-day percent. Take a piece of paper that is

2 days wide; you slide it back and forth along the histogram until you find the pair of adjacent days that gives you the largest probability, as in Figure 18.1. So a particular origin-destination pair may have a 2-day percent of 85%, which means that the percentage of travel time in the best contiguous 2 days, in this case 3 and 4 days, was 85%. In the case of a high-quality intermodal service, the 2-day percent might well be 100%.

This measure is manipulable as well; one could use a 3-day percent, rather than 2-day percent, and change the apparent performance. But, it does have the ability to measure central tendency in a useful way. If the customers know where that 2-day percent is on the distribution, they can gear their operation toward delivery in that time window with some degree of confidence. It also eliminates the impact of bad data points on the right tail. This is a useful measure of performance in the railroad industry [1].

Other Rail Issues

To round out the discussion of railroads, we briefly touch on various rail issues—many of these have parallels in other modes; these issues are illustrative of concepts inherent in freight transportation.

....

Empty Freight Car Distribution

Empty freight car distribution is a fundamental concept. We often have unbalanced flows and have to redistribute empty freight cars to loading points. In the United States rail industry, about 58% of the freight car-miles are loaded.

If the railroad is fortunate, it finds a load right at the spot where the car was made empty by the receiver of goods. More typically, there is a triangular empty move to spot the car for loading at another point in the system, as shown in Figure 18.2.

How one goes about redistributing empty cars on the system is important. There are costs associated with dragging empty cars around. It uses power; it uses capacity; it is not free. So you want to reposition cars where you think there might be loads; are you better off leaving the car where it is until a load materializes—or are you better off moving

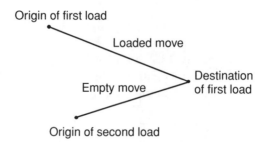

FIGURE 18.2   *Empty moves.*

it to another point on the network on speculation? In practice, we know a lot about where the demands are on the network, so we have standard empty moves taking place. Remember—getting those empties to demands quickly is important. It is a level-of-service variable. A properly repositioned empty car generates revenues.

Factors in Movements of Empty Cars

There are some complicating factors in the movements of empty cars on the rail network. Here we present several of these factors that stem from the institutional structure of the United States rail industry.

Car Ownership

In the rail industry freight cars are owned by different railroads, as well as private companies which provide freight cars to the industry at large. The Union Pacific owns some cars, the Norfolk Southern owns some cars, and so forth. But the cars flow freely on the network. That is, Union Pacific does not usually insist that its cars remain only on its track. Consider a car loaded on the Union Pacific for a destination on the Norfolk Southern.

The Union Pacific-owned car goes onto the Norfolk Southern lines and is now a "foreign car." What happens when it is made empty?

The Norfolk Southern can reload those Union Pacific cars, but only for a destination heading back towards the Union Pacific. The Norfolk Southern could load one of its own freight cars—called a system car—for any destination it wanted to. So there are "car service" rules associated with what one can do with empty equipment, depending on who owns that equipment (see Figure 18.3).

FIGURE 18.3   *Foreign*
empty moves.

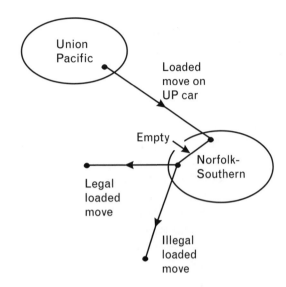

Per Diem

There is a concept in the railroad industry called "per diem." Basically per diem is a rate that railroads pay each other for the use of equipment. It is a daily rate that is supposed to reflect the cost of owning the car. If that Union Pacific car is on the Norfolk Southern, either loaded or empty, the Norfolk Southern pays the Union Pacific to rent that car.

That leads to all sorts of interesting questions. Railroads do not want to have foreign cars on their property for which there is not a load, because they pay for those cars. At the end of the month, funds are transferred reflecting how many car-days they rented from each other this month. It is a cash outlay.

Suppose we have a car surplus; there are more empty cars than there are loads. There may be a number of Norfolk Southern empties on Union Pacific; there may be a number of Union Pacific empties on Norfolk Southern. In the time of car surplus, Union Pacific says, "I want to get these cars off my property; it is costing me $15 a day a car. I do not need them," and Norfolk Southern is saying exactly the same thing. So you will have 100 empty cars going from Union Pacific to Norfolk Southern, and that movement is not free. And you will have 100 empty cars going from Norfolk Southern back to Union Pacific; that movement is not free either. Now, if one looked at this in a more global sense, the obvious answer is for these two railroads to say, "Look,

I will hold your cars; you hold mine; neither of us will have to pay to move these empties around."

Clearing House

In fact, some railroads have occasionally developed this as a formal relationship called the "clearing house," composed of a number of railroads. For the purposes of empty car distribution, they treat all cars from those railroads as system cars.

How well clearing house operations work depends on demand. When the economy booms, everybody needs cars for their shippers. If the Union Pacific has fewer cars owned by the other railroads on their system than there are Union Pacific cars off their system, they will be unhappy. So the clearing house concept, which works well in times of car surplus because we avoid extraneous empty moves, works less effectively in times of car shortage, when everyone wants cars for their shippers.

Interline Moves

Railroads are independent private carriers. A shipment may originate on one railroad and terminate on another railroad. So Union Pacific may originate the traffic, and Norfolk Southern may terminate the traffic. This is called an "interline" move. There is the need for some cooperation in providing this interline service to the shipper. All the shipper cares about is the origin-destination level-of-service; whether it is the Union Pacific or Norfolk Southern that is at fault for a poor move is of no concern to the shipper.

The Union Pacific and Norfolk Southern need to be carefully coordinated at interchange points, which often are sources of trouble. When traffic moves from one railroad to another, service quality tends to deteriorate.

So you have a strange business relationship that can arise; railroads, on the one hand, are supposed to be cooperative entities providing service to a shipper that requires both their networks; on the other hand, these railroads are competing vigorously for some other shipper's business. So railroads deal with that kind of conundrum—cooperation as well as competition (see Figure 18.4).

FIGURE 18.4
*Competition and
cooperation.*

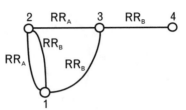

RR$_A$ and RR$_B$ cooperate on 2–4 traffic
and compete on 1–2 traffic

....
Mergers

Mergers have been an important factor in the United States rail industry for decades. This industry restructures itself through that mechanism, going back to the last century. Currently, the major rail carriers are structuring themselves into a few, very large railroads.

Parallel and End-to-End Mergers

There are fundamentally two kinds of rail mergers (see Figure 18.5). There are parallel mergers between companies that provide services in the same corridor; in parallel mergers, the railroads would say, "We are both losing money on this corridor; there isn't enough business for two railroads and the costs of maintaining two rights-of-way is prohibitive. So, let us merge so one company can survive."

In parallel mergers, the obvious concern is a lessening of competition. Shippers want to have a choice between railroads so that "rate discipline" occurs through competition.

There are also end-to-end mergers. Here two railroads that do not compete say, "We can together provide a much better service merged than we can individually. This transfer of freight cars becomes an intra-company, rather than an inter-company transfer, so service will improve and cost savings will occur as well."

Federal regulators will often look positively on end-to-end mergers, because they view them as a way of attaining cost savings and better services to the customer, without reduction in competition.

Suppose railroad A and railroad B merge as shown in Figure 18.5. There is a corridor in which they compete; there is a corridor in which they are end-to-end. Very often mergers have both parallel and end-to-end characteristics.

FIGURE 18.5   *Types of mergers.*

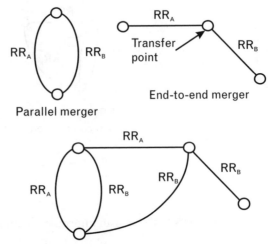

Parallel merger

End-to-end merger

Combined parallel/end-to-end merger

....

Power Distribution

We next turn to power distribution on railroad networks, which is a complex task. You may need a certain number of locomotive units to pull a train. There is some flexibility depending on the operating speeds you want to maintain and "ruling grade," which is simply the steepest hill you have to climb, going between point A and point B.

The idea is to ensure power is in the right place at the right time. One does not want to "deadhead" power, if at all possible. When locomotives pull a train, they are performing a revenue-generating service, but they are also being repositioned for another train. So, if a train is dispatched late and therefore arrives late, or is canceled altogether, we are affecting the level-of-service of cars on that train, of course. But, the level-of-service on the outbound train that the power from the incoming train will be assigned to is negatively affected as well. It is a complex systems problem.

....

Maintenance

Maintenance is an important concept in transportation systems. Rail-roads maintain the rolling stock, power, and the right-of-way. Doing preventive maintenance, as opposed to demand-responsive mainte-nance—fixing something when it fails—is the desired procedure.

There is an important relationship between the level-of-service and the quality of the physical assets. Sometimes, maintenance will be deferred. For example, if cash is short, it is easy to say, "This is a tough financial year; maintenance on the line from Albany to Buffalo can wait until next year. We won't be compromising safety." When maintenance is deferred, deterioration of the physical facilities accelerates rapidly. Catching up can be difficult for railroads (and other modes).

Types of Service

Generally speaking, in the rail industry there are three kinds of services.

General Merchandise Service

First there is general merchandise service. The network in Figure 18.6 is an example of general merchandise service.

Intermodal Service

Next, there is intermodal service. Here cargo is loaded into containers that are transported by rail on flat cars. These are intermodal in the sense that those same containers can be transported by truck, or can go

FIGURE 18.6   *Rail network: merchandise service.*

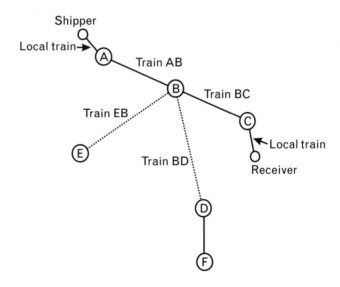

internationally via ship. The railroads are developing partnerships with trucking companies; they provide the long-haul services for trucking companies, who provide pick-up and delivery service. Also, partnerships among railroads, trucking companies, and ocean shipping companies have proven useful from a service and cost perspective.

While a 2-day percent measure is used for general merchandise service, for intermodal traffic, the appropriate measure may be a 6-hr percent. In merchandise, we deal in days. In intermodal, we deal with hours. Plus or minus 2 hours is a reasonable kind of service target for intermodal. This is a wholly different level-of-service, since the customers are used to truck level-of-service.

Bulk Commodities

Finally, railroads transport bulk commodities, such as coal and grain. These are often transported by unit trains carrying only that one commodity; so you have a unit coal train, which typically goes from origin to destination loaded, and from destination back to that same origin empty for reloading. These commodities are very sensitive to transportation rates, because these rates are a large fraction of their overall value. If you are transporting coal, a few cents on the freight rate means a lot, relative to the value of the coal. If you are transporting television sets, a few cents on the freight rate is a relatively modest fraction of the overall cost of the item. The dominant variable in level-of-service for bulk commodities is the rate they are charged. Travel time and reliability are much less important.

Safety

There are several aspects of rail safety. Railroads transport hazardous materials. When cars derail and hazardous material escapes, it can be a very serious situation, particularly if it happened in a populated area. So "HAZMAT," as it is called, needs to be properly transported.

Grade-crossing safety is an important issue. There are on the order of about 300,000 grade crossings in the United States—places where automobile and/or truck traffic crosses at grade with a railroad right-of-way. About 600 people a year over the last several years have been killed at grade-crossing accidents in the United States. That represents about 1.5% of the total highway fatalities in this country.

Another safety-related area is control systems such as automatic train separation (ATS) or its more advanced cousin, advanced train control (ATC). Both are designed to minimize train collisions. The question is cost versus the safety one achieves with various configurations of ATS and ATC systems. This is a difficult public policy question. The private railroads and the Federal Railroad Administration (FRA, of the U.S. Department of Transportation) often differ on this topic. The railroads feel that the major capital investment required for those systems, if mandated by the FRA, are not commensurate with the achievable safety improvements.

Final Comments on Rail

Using different levels-of-service as a way of managing capacity is an important transportation concept, and is applicable to railroads. There are many different kinds of capacities to be allocated. There is line-haul capacity—a track can accommodate only so many trains a day; there is terminal capacity, as one switches trains through those facilities; capacity is also affected by the number of freight cars in our inventory, the amount of power on the system and labor and management resources, among other items. Earlier, we described Kwon's simple two-node model and how one managed capacity measured in terms of empty-car inventory and power availability.

The amount of flexibility in rail operations is an important design question, particularly when one operates close to capacity. This is being tested now, because the railroad industry is at historically high levels of traffic while trackage has been reduced substantially. Articles in the professional press stress the line-haul capacity problems of railroads, which has not been an industry issue for decades. Traditionally, the problem has been excess capacity and the need to rationalize. Now railroads are in the situation in which there is not enough capacity to provide the services they want to provide; ideas, concepts, and operating policies that worked well at 50% of capacity are not working very well at 95% of capacity. How well the industry addresses this issue will govern its success in the future.

REFERENCE

1. Martland, C., P. Little, and J. Sussman, "Service Management in the Rail Industry," Proceedings, AAR/FRA/TRB Conference on Railroad Freight Transportation Research Needs, Bethesda, MD, July 1993.

CHAPTER 19

Trucking

We now discuss several other freight modes: trucking, ocean shipping, and intermodal transportation.

Trucking

Trucks differ from railroads in a number of fundamental ways.

In the United States, trucks differ in right-of-way ownership. Freight railroads in the United States run on right-of-way that they own and maintain. It is mostly a fixed cost. The trucking industry uses the public highways, which, in general, they pay for on an as-used basis—it is a variable cost.

The trucking industry is much less concentrated than the railroad industry in terms of the size of firms relative to the overall size of the industry. The railroad industry has gross revenues of over $30 billion per year. The four largest firms in the rail industry have upwards of 80% of the total revenue.

But the giants of the trucking industry (e.g., Yellow Freight and Consolidated) when you add their revenues together, are still a relatively modest fraction of the industry. It is a much more diffuse industry with many more firms, and many very small firms. However, its gross revenues of $280 billion dwarf those of the United States rail industry. Since rail's market share in ton-miles is larger than that of truck, clearly the trucking industry has much higher rates, reflecting the higher value freight they carry.

The two modes differ in right-of-way technology. The railroads are steel-wheel on steel-rail; the highways are rubber tires on concrete or asphalt. The power sources are different. In trucking we use internal combustion or diesel engines; in railroads, we use electric or diesel power. In railroads, we have rail cars that are unpowered, pulled in trains of multiple vehicles by locomotives. Trucks have tractors that usually pull a single trailer, but over the past several decades, the trucking industry in the United States has begun to run what are called "tandem-trailers" or "double-bottoms," two trailers in tow.

There is a good deal of debate between the railroad and trucking industries about the safety of this configuration. The railroad industry says this is an unsafe mode; they think it is a public menace to have these double-bottomed trailers on the nation's highways. The trucking industry likes the productivity gains that they get by pulling two rather than one trailer—exactly like the railroad industry's productivity gains by having double-stacked container trains rather than single container on flatcar operations. The trucking industry, of course, argues that double-bottoms are perfectly safe and that it is positive from a societal viewpoint since the costs of transportation are lower and everybody benefits. As always, "where you stand depends on where you sit."

In effect, when one starts to hook trailers together, you have pseudo-train operations. It is not steel-wheel on steel-rail, but you do have a tandem operation that, in operating effect, is a short train.

Trucking Cost Structure

As noted above, unlike railroads, the motor-carrier industry tends to be primarily a variable-cost rather than a fixed-cost industry. This is not surprising, given the fact that they do not own their own right-of-way. They pay through taxes for their right-of-way use. They use less of it if business is down and they pay less. If they use more of it because traffic is up, they pay more. If you own the right-of-way, as do the railroads, you pay for it every day regardless of use.

Truckload Operation (TL)

There are basically two operating modes in the trucking industry. The first is truckload shipping, abbreviated TL. This is, in concept, a straightforward operation. A trucker is in the business of providing and driving a truck to take your goods from point A to point B. You, in effect, rent the truck for that move; the driver drives the shipment from A to B. This is an origin-destination service; the truck is dedicated during that trip to a single shipper. The truck goes from A to B; you pay your money and the trucker looks for another load.

The driver may be an independent owner/operator who owns, perhaps, only that one truck and who is desirous of keeping it as productive as possible. To help the owner/operator, there are services that provide information about potential loads. And if that new load is right up the street from where he dropped off the earlier load, then there is no dead-head time. But if he needs to travel some distance to pick up a new load, no one pays him while he does that.

In addition to the small owner/operator, there are some very large TL firms as well. Schneider and J. B. Hunt are examples of very large truckload firms. They strive for the economies inherent in large fleets: better fleet management for better productivity, coordinated maintenance, driver training, and so forth.

Load-Screening

We now look in more detail at the decision by a trucker, given an opportunity to take a load, as to whether or not he should do so. This is called "load-screening." Suppose we have an owner/operator who is offered the chance to take a load from a particular origin to some destination (see Figure 19.1).

How does the trucker make a decision? The trucker must consider the revenues that he will generate from the new load. Further, he must consider the costs of taking the load from origin to destination. Then, he has to consider the value of being at the new position—the destination of the new load. What are his chances of getting another new load somewhere in that geographic area?

Further, the trucker has to decide whether to move empty from his current position to the new load or whether a better strategy would

FIGURE 19.1
Choosing a load.

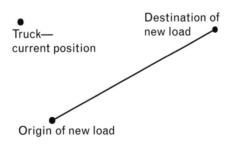

simply be to wait where he was unloaded for a new load that he can get with no dead-head move.

This is a complex and probabilistic system. Loads are being generated randomly in time and space. There are many different kinds of loads with different revenue and cost characteristics.

Now, consider this from the point of view of a company that has a large fleet of trucks. This company also has many demands for service; while these demands are random, the large shipper can take advantage of the scale of the operation to reduce some of the stochastic elements.

The large company goes through a process of load-screening—whether or not to accept the load; dispatching—which truck to assign to the load; pricing—considering the level of competition, as well as the relationship with a particular customer, to determine a price for the move.

Large truckers have developed models where they assign a value for a truck at a particular node in their network; they then consider the fleet management problem as one of trying to assure that trucks are at "valuable" nodes in the network—nodes at which good loads are available, where "good" is determined by the cost and revenue characteristics of likely loads at that node and where those loads will take the truck (hoping for another "good" node) (see Figure 19.2).

As you would expect, information technology helps this kind of optimization. The trucking companies are dealing with real-time decision-making with a great deal of information available about both supply of trucks and demand for service.

We note that the vehicle cycle, which we discussed for railroads and elevators, is a fundamental element in trucking as well. The truck is the key asset. Keeping it productive is vital.

TL Markets

TL tends to be a rough-and-tumble market, particularly for the owner/ operators. These are people—entrepreneurial, typically—who are often

FIGURE 19.2 *Empty truck positioning and dispatching.*

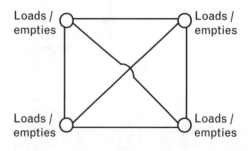

running on a shoestring. Perhaps they have mortgaged all their assets to buy a truck. And now they try to generate enough revenue to make a living and support their families; clearly they will price aggressively and work to maximize productivity.

The owner-operators provide a very important service. It is a flexible, responsive system with all these independent people trying to get loads; they do it in a very efficient way, sometimes with cutthroat price competition, which helps shippers. They provide a ubiquitous productive service around the country.

Intermodal Partnerships

The large truckload firms provide services for containerized freight in the intermodal market. They will often pick up a container from a shipper and bring it to an intermodal rail connection. For example, J. B. Hunt is a large TL carrier; Conrail is a major railroad (which in June 1999 split into two portions owned by the Norfolk Southern and CSX). They have had a very close partnership providing intermodal services. J. B. Hunt provides service from the shipper to a rail yard. Conrail provides a long line-haul service, and then J. B. Hunt picks the container up and provides truck service to the ultimate destination (see Figure 19.3).

These two disparate companies, a railroad company and a trucking company, have reached an accommodation using their modal advantage. J. B. Hunt is very good at accessing shippers and receivers—all you need is a highway. Conrail is good at providing cheap long-haul service. It is a natural partnership. They take advantage of the ubiquitous highway network and the low railroad costs on the long haul.

The challenge for the railroad industry has been that customers (shippers and receivers) who use trucks are used to and expect high-quality service. They expect high reliability. So the railroad industry

FIGURE 19.3
Intermodal partnership.

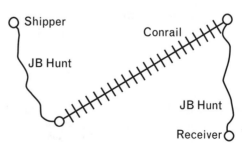

cannot treat these containers just like any shipment. It must treat them as though it is a truck that happens to be on a railroad at this particular moment. The kinds of service that the railroads provide for intermodal reflect the high-quality service that trucking customers expect.

Some railroad people grumble about the fact that, "When those intermodal trains come through, everything has got to get out of the way. Even if we have an important merchandise train, they're always giving those intermodal trains all the priority, because of the very tight service windows that have been promised." There are those in the railroad industry who feel that, although there are very good rates of growth in intermodal traffic, they are not making very much profit on intermodal. They suggest that because the services that are provided are very expensive, rates are such that it is not a moneymaker.

The fundamental challenge of intermodal transportation is as follows: you use the inherent advantages of each modal partner—the universality of the highway/truck network and the low-cost line-haul attribute of the rail network. But if you cannot do an efficient transfer between the two of them, you dissipate the advantage. Containerization is a fundamental aspect of that. The fact that containers are uniform in size and transfer equipment (e.g., cranes) exists to deal with containers—moving them readily from one mode to the other—is fundamental to the idea of intermodal transportation.

This technology has allowed intermodalism to develop. Further, information technology and advanced sensors, allowing for coordination between modes and for so-called "in-transit visability"—knowing where your shipment is at all times—have enabled more rapid deployment of modern intermodalism.

Less-Than-Truckload Operation (LTL)[1]

The other truck operating mode is "less-than-truckload"—LTL—which in some ways is a wholly different industry from TL. They operate more like a railroad than a truckload shipper.

---

1 An excellent reference on LTL operation is Caplice, Christopher G., "An Optimization-Based Bidding Process: A New Framework for Shipper-Carrier Relationships," Chapter 2, Ph.D. Thesis, Department of Civil and Environmental Engineering, MIT, June 1996.

LTL Networks

LTL trucking companies have terminals and a feeder network that picks up and delivers freight from shippers in small trucks and brings that freight to end-of-line terminals; there those shipments go into a line-haul truck that travels between terminals, at which point the reverse operation takes place. The shipments are disaggregated and are distributed, typically, in smaller trucks (see Figure 19.4).

This is not unlike a railroad system. The feeder truck routes are like the branch lines in the rail network, where a local train goes out and drops off some empties, and picks up some loaded cars to bring them back into the main system. Shipments are consolidated for the line-haul move from a major terminal to the next, just as freight cars with the same destination are consolidated in trains. Understand that those line-haul trucks have my shipment and your shipment in it. We are bundled together, for our mutual advantage, unlike the truckload operation, where the entire cargo is usually for one shipper/receiver pair.

FIGURE 19.4
LTL network.

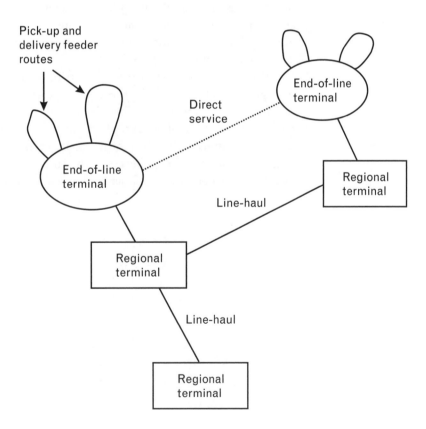

LTL Carriers

There are big firms in the LTL market: Yellow Freight, Consolidated Freightways, and Roadway are the three giants of the industry. These LTL truckers have a capital structure that is very different from truck-load carriers. Truckload carriers have no consolidation terminals, since they do not deal with the consolidation process. On the other hand, LTL truckers have a big capital structure. They have real estate for terminals; they have substantial administrative costs for dealing with routing and scheduling.

Regional Versus National LTL

There are several different kinds of LTL carriers. Regional carriers, who serve the Northeast, the Southwest, and so forth, are one type. Also, we have national LTL carriers that provide coast-to-coast service under one administrative umbrella. We have seen the development of alliances of regional carriers that coordinate their operation to provide long-haul services to customers who have interregional flows of traffic. This is shown in Figure 19.5.

There is competition between the national LTLs and the alliances of regional LTLs. Typically the national carriers are unionized; the regional carriers are non-union. This tends to give the alliances of regional carriers a cost advantage. Another trend in the industry that favors the regional rather than the national carriers is that length-of-haul is getting shorter as suppliers—in this highly competitive market-place—tend to be locating closer to their customers.

FIGURE 19.5
Interregional LTL.

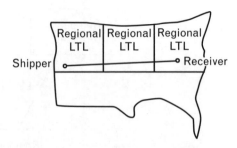

New Trucking Technologies

The trucking industry has been introducing various new technologies that are of value in improving operations and service. Many of these technologies are part of intelligent transportation systems (ITS) (see Chapter 24).

Automatic Vehicle Location

Automatic vehicle location (AVL) locates all vehicles in a fleet, be it an urban fleet like the UPS package cars, or an intercity fleet, like Yellow Freight. Global positioning systems (GPS) are often used, although roadside infrastructure can be used as well. The infrastructure senses (i.e., "reads") a transponder on the vehicle, which has a unique identifier, locates that vehicle, and passes that information to a control center. The ability to locate and identify trucks allows us to do a better job of routing and scheduling those vehicles, thereby making more productive use of the vehicle fleet. Also, keeping track of where those vehicles are has some security advantage.

The trucking unions are very concerned about privacy implications of this technology. As with the introduction of many new technologies, there are some labor relations issues.

In-Transit Visibility

These location systems can be a mechanism for providing information in real-time about shipment location to shippers. Some technologies involve transponders on individual shipments in order to track them. One places these transponders on containers that can be sensed remotely; as that container goes from truck to rail to ship, one can continue to track the shipment. This is called "in-transit visibility."

Weigh-in-Motion

Another new technology in truck transportation is so-called "weigh-in-motion" (WIM) devices. In order to properly tax trucks, you have to know how much they weigh. The weigh-in-motion technology permits the weighing of the truck with in-ground sensors at highway speed; the truck does not have to stop at a weigh station, thereby compromising productivity.

There are laws that limit the weight of trucks. There are three issues. Safety is one. A second is taxation. Taxes depend on weight and distance. Third, overweight trucks may cause excessive damage to the highway.

Paperless Transactions

Another trucking technology concept is paperless transactions. The various political boundaries between Mexico and the United States, Canada and the United States, even between the various states, requires substantial paperwork and documentation. Doing this in a paperless way through technology is a tremendous cost advantage.

Private Carriage

Private carriage is another aspect of trucking. There are some companies that provide for their own truck transportation needs because of their service requirements. So they own trucks and hire their own drivers to move their goods. Many companies will have a private fleet, but they will use TL or LTL occasionally.

That concludes the discussion of trucking. Ocean shipping is the next topic.

CHAPTER 20

Ocean Shipping, International Freight, and Freight Summary

Ocean Shipping

The world's economy is increasingly global. Most of us sense this quite directly through communications technologies such as telecommunications, faxes, and e-mail; people are able to talk to colleagues in Japan, Bangkok, and France, as easily as Arlington, Massachusetts. This is commonplace. It is a very small high-tech and international world.

But there is a lot of conventional transportation moving goods around this global economy. The overwhelming proportion of international shipments are by ship on ocean-going vessels.

These ships are enormous vehicles compared with railroad freight cars or trucks. They are relatively slow. A fast ship goes on the order of 20 knots, or nominally 20 mph, which by railroad train or truck standards would be slow indeed. And it tends to be relatively inexpensive per unit weight and per unit distance to transport goods by ship relative to other ways of traversing the ocean; certainly, it is much cheaper than air transportation, and cheaper on a ton-mile basis compared with surface transportation modes like rail or truck. About 5% of the world's freight bill is for ocean shipping.

Ocean Shipping Services: Wet Bulk and Dry Bulk

As with truck and rail, there are different kinds of services in ocean shipping. Bulk commodities are of two kinds: wet and dry bulk. Wet bulk is mostly oil products, which are shipped in the holds of gigantic oil tankers. We have dry bulk as well—iron ore, coal, grain, bauxite, aluminum, phosphate. These ships operate like truckload trucks. They travel from origin to destination. They are unloaded and travel empty to pick up another load.

The bulk market has some similarities to the TL market in the trucking industry. It is a very volatile market; the freight rates and demand for service may swing widely over the course of a few days. It is a very easy business to enter and exit in the same way that truckload trucking is; you buy a truck and you are in business, you buy a ship and you are in business. So you can easily get in and out, and that leads to the chronic over-capacity and volatility in that marketplace.

Demand may vary substantially over time. There are often major peaks and valleys in demand. There is enormous competition on price. Here, we are carrying commodities where the transportation cost is an important fraction of the overall cost.

An empty vessel is not earning any money for the owner, so we see very aggressive pricing taking place when there is a surplus of ships. Pricing service at or even below operating cost—actually losing money on the operation to keep your crew together—often occurs.

These ships are enormous; economies of scale exist in ocean shipping as they do in most transportation services (for example, long trains, double-stack containers, double-bottom trailers, and so forth).

Environmental Issues and Risk Assessment

The sheer size of these vessels has environmental consequences. An oil tanker ran aground in Prince William Sound in Alaska several years ago, spilling thousands of gallons of oil in an environmentally sensitive area; the environmental impacts and cash settlements were enormous.

How does one go about avoiding these environmental risks in ocean shipping? Double-hulling the ship is one approach. If you run aground and puncture the first hull, you still have a second hull that is containing the oil. You can, of course, puncture both hulls if the crash is substantial enough. Needless to say, it costs more to build a double-hull ship than it does a single-hull ship; ship owners may or may not be motivated to do so, depending upon the risk/cost structure. Usually, they have to be induced to make these kinds of choices through regulation.

Crew training is an important issue in risk mitigation. Usually, the ships are highly automated, so the numbers of people you actually need to run it is small. But the training for those people is very important in avoiding accidents. Investments in aids to navigation, usually a public-sector investment, is important as well, not unlike air traffic control in the airline industry.

There is a relationship between speed and safety. If the ship is going fast and collides with something, the kinetic energy is higher and a fracturing of the hull is more likely. Also, the probability of something happening at all is higher because there is less time to react. You cannot turn these tankers very quickly.

The regulatory aspect of safety—the interaction between the private and public sector—is important. Ship owners want ships that are cheap to build and run them at top speed with minimal crews; competition in this cost-driven industry gives them the incentive to act that way. The regulators want well-built expensive ships, say double-hull ships, operated at reasonable speeds with well-trained and redundant crews. You have these two opposing perspectives. The general public wants to split the difference. We want a clean environment—the public did not like it when the Prince William Sound accident occurred. But at the same time we want cheap oil, and if it costs more to transport oil because of all of these regulations, ultimately we pay. So the political process finds a point between these two extremes, trading off the benefit of cheap oil with the benefits of a clean environment.

The Liner Trade

The other shipping operation is the liner trade. Usually, for merchandise (as opposed to bulk), the cargo that an individual customer has is not adequate to fill a ship. So, in the liner trade, as in LTL operations and in merchandise trains, a customer shares the capacity with other customers, and benefits from consolidation by sharing the costs as well.

Loading and unloading times in such operations are very important. Ships do not make money when they are sitting at the port being loaded and unloaded. The development of containers was motivated in part by this consideration. You put merchandise in containers; it is a standard size; it can be loaded on and off the ship much more expeditiously, and that big, expensive ship gets turned around very quickly.

Liners have pick-up and delivery at multiple ports, as with LTL operations. For example, consider Figure 20.1.

The liner travels from A to B in the United States; it then crosses the Atlantic and goes to Ports C, D, and E and then returns to Port A in

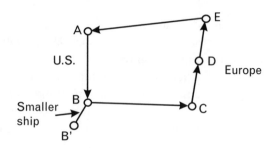

FIGURE 20.1 *Liner operation.*

the United States, picking up and dropping off shipments along the way. In addition to the pick-up and delivery occurring at each of these ports of call, there may be what is called "trans-shipping." At Port B there may be some off-loading of some containers or some general merchandise that goes on some other smaller ship or possibly even a barge, to some other Port B'.

Containerization

On an international scale, about 60% of nonbulk cargo is currently containerized. Shipments either originating in or destined for the United States are about 80% containerized.

The liner market, unlike the bulk market, is a stable operating environment. Schedules are published; frequencies are fixed for six months in advance; shippers have long-term contracts with liner companies; it is a much more organized market. It is comparable to LTL trucking, which is a more organized market than TL.

Conferences

Liner companies often form what are called "conferences," which basically set prices; they are legal price-fixing organizations. This provides stability to the liner companies. Also, shippers have guaranteed rates into the future, albeit perhaps a higher rate than they might have in a totally free market environment. What shippers obtain by allowing the liner carriers to collude in setting prices is a relatively narrow guaranteed window of rates, but at a higher rate level. In a free market system, they might get a lower rate on average, but the volatility of that rate would be higher.

Liner Decisions

Liners have multiple origins and destinations, and many customers; it is the classic network problem of serving those multiple origins-destinations and multiple customers with their different service needs. What kinds of decisions do liner companies have to make? Which ports they should serve—the Port of Boston or the Port of New York or both? What ships to purchase? They typically have a fleet of ships with different capacities and operating characteristics. What ship should a company use on a particular route?

Economies of Scale

As with bulk commodities, we have economies of scale in ship size. If one plots the cost per container for transportation as a function of the size of the ship, it is as shown in Figure 20.2.

The bigger the ship, the lower the cost per container. So there is an advantage on the cost side to have large ships, just as there is an advantage for railroads to operate long trains. But there are some disadvantages; the quality of service is not as good. Perhaps with very large ships, service is infrequent. Certainly the service is not as nimble and as responsive as it might be with smaller ships. So, again, we have the cost-service trade-off.

Operating Speed and Cost

Another issue that the liner company deals with is operating speeds: how fast do you run the ships? There are trade-offs in terms of the quality of service offered and cost. Ships burn fuel faster, as you might expect, when they travel faster, just like your automobile does. Moving up from 18 to 20 knots has more than a 10% impact on your fuel use—there are some nonlinearities in that relationship.

FIGURE 20.2
Economies of scale in shipping.

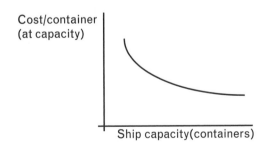

Service Frequency

Liners typically provide much less frequent service than land-based modes. The time frames are weekly service, typically, or twice-a-week service, or perhaps once-every-other-week service.

Empty Repositioning of Containers

As with the other modes, there is an empty repositioning question—one has to move empty containers to places where they can be loaded again. Ships are carrying some percentage of empty containers. About 20% of worldwide container moves are empty. Now a rule of thumb used in the shipping industry: a company providing liner services needs about 2.4 times the container capacity of its fleet to provide adequate service.

Intermodalism and International Freight Flows

The key driving force behind containerization, in addition to efficiency at the ports, is providing intermodal transportation service. We illustrate this using a stylized United States and Europe as shown in Figure 20.3.

The intermodal operation consists of a truck in the United States to a rail link, to a port, to a ship; then we reverse the process: a rail link in

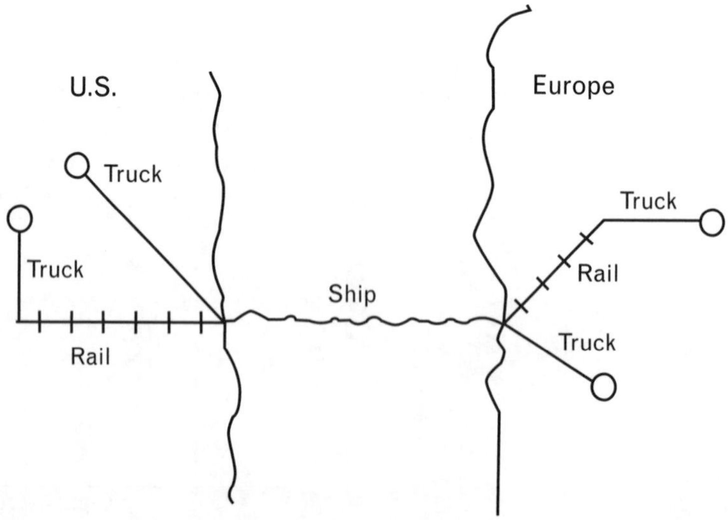

FIGURE 20.3
U.S./Europe intermodal services.

Europe to a truck to the ultimate destination. The container retains its integrity throughout. The shipper loads and seals that container in the United States; the receiver opens it up in Europe.

There could be the need for some additional trucking moves. Sometimes a truck move from the port to the railhead is needed; rail access to the port may not exist. So you may have to put that container on a truck to access the European rail system in this case, and you may have to do the same thing in the United States.

Some companies broker container services, like LTL trucking, combining loads from different shippers to fill a container, thereby gaining economies of scale. So, when the container comes into the port warehouse, a trucking company picks up its portion of the shipment from the container and makes its delivery.

The idea of intermodalism is that each mode is used as effectively as possible; we have that ubiquitous highway/truck network; we have the rail service at low cost for long-haul; and we have ships to cross the ocean.

There are many international shipping patterns that have developed. Here is a representation of some of these patterns (see Figure 20.4).

International Trade Patterns

Various services link together Europe, Asia, and the United States. Shipments from Europe to the West Coast of the United States can go across the Atlantic Ocean, through the Panama Canal, and back up to, say, the Port of Long Beach or the Port of Los Angeles. The other route is the "mini land-bridge," where a shipment goes from Europe to the East Coast of the United States and on to the West Coast by rail. Similarly, to go from Asia to the East Coast, one can go through the Panama Canal, and to the East Coast of the United States. That same mini land-bridge concept also is used with shipments going from Asia to the West Coast of the United States by container ship and by rail—the mini land-bridge—to the heartland of the United States and to the East Coast.

In some instances a land-bridge concept is used to get from Asia to Europe: Asia to the United States by container ship, across the United States by land-bridge via rail, and on to Europe on another container ship. In some cases, this is more efficient than going through the Panama Canal from Asia to Europe.

FIGURE 20.4
*International trade
patterns.*

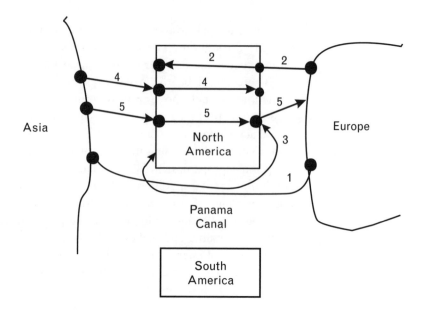

1) Europe–U.S. (West Coast) via Panama Canal
2) Europe–U.S. (West Coast) via mini-land bridge
3) Asia–U.S. (East Coast) via Panama Canal
4) Asia–U.S. (East Coast) via mini-land bridge
5) Asia–Europe via land bridge

Manufacturing in Asia is moving west from Japan and Korea into Southeast Asia in countries like Thailand and India. Formerly, the manufactured goods went from Japan and Korea to the West Coast ports of the United States. Another way of handling the traffic, now that the manufacturing activity is taking place in Southeast Asia rather than in East Asia, is to go through the Suez Canal and the Mediterranean Sea out to the Atlantic Ocean and to the East Coast ports of the United States. So the flow of manufactured goods from the Pacific Rim, as the manufacturing shifts from Korea and Japan to Southeast Asia, has started now to favor the Atlantic Coast ports rather than the Pacific Coast ports (see Figure 20.5). This raises interesting questions about the competitiveness and productivity of the East Coast ports as compared to the ports on the West Coast and investments in these ports as regional and national strategies.

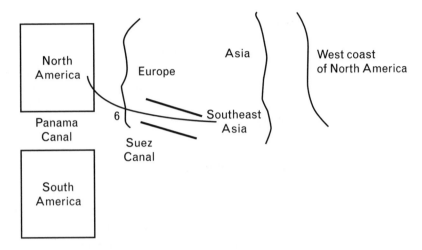

FIGURE 20.5 *More international trade patterns.*

6) All water route from Southeast Asia to east coast of U.S.

Port Operations

For intermodalism to work, port transfers have to be efficient. This is an integrated system; the customer thinks of it not as a set of individual modes, but as a single integrated service. Efficiency within the ports is important to the process.

These ports are a complex of dock-side cranes that can rapidly load and unload the containers from ships. Land-side access for rail and truck is needed to pick up and deliver containers. Storage room for containers is needed as well.

These ports are typically government-owned, or owned by quasi-public organizations like Massport or the Port of New York and New Jersey. There is enormous competition among ports—ports fight vigorously for traffic because it has important regional economic implications.

Port Capacity

Port capacity is an important issue. How many ships can be there at any one time? You do not want expensive ships queuing for unloading berths at a port.

Dredging

Another interesting issue is dredging, which is simply removing the dirt from the bottom of the harbor so that large ships are able to access the port. Dredging is a continuing operation, as the harbor bottom shifts naturally. However, there are environmental questions. The bottom contains toxic material in some areas, which is disturbed by dredging.

Intermodal Productivity

The trend in the industry is toward larger ships to achieve economies of scale. So, port productivity becomes more important; those ships have to be loaded and unloaded expeditiously.

All this is part of an emphasis on increased intermodal capacity. Intermodalism is certainly a big and growing business, with bigger ships, double-stack technology on the railroad, bigger containers, bigger container ports, more efficient and higher-capacity cranes. For increased productivity, port automation is important as well. Containers may have transponders on them for tracking by GPS—Global Positioning System. Faster container ships improve productivity as well.

The Total Transportation Company

There is vertical integration taking place in this industry, creating total transportation companies, providing intermodal service in an integrated way. Further, specialized companies are now providing total logistic services. Shippers outsource their entire transportation function to these specialists, who then provide for all the shippers' transportation needs.

Information Technology

Efficient transfer between modes is a central concept for intermodalism. Information technology is essential to organizing and expediting these intermodal moves. While international freight flows drive the intermodal marketplace, information technology enables it.

Information technology is used in various ways: pricing and revenue collection, which includes electronic data interchange (EDI); information services such as schedule information to customers in real-time; in-transit visibility; operations control, which includes real-time service management; connections between modes and terminal operations; and management information, which includes revenue allocation between intermodal partners, performance measurement, and so forth.

Some carriers emphasize moving information as a fundamental service. Bram Johnson of Roadway Package System was quoted in *Traffic World* saying, "Today, we are not in the transportation business, we are in the information business." And he goes on to say, "You must be able to move information in advance of the box."

Dr. Aviva Brecher, of the Volpe Center, developed a chart that shows the various intermodal freight technologies and developments on the facilities and equipment side, as well as the information systems side. From that, we get a sense of the broad range of technological activities supporting the intermodal revolution (see Figure 20.6).

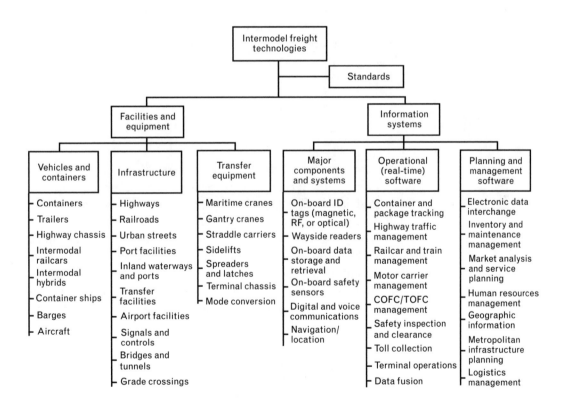

FIGURE 20.6   *Intermodal freight technologies. (Source: Dr. Aviva Brecher, Volpe National Transportation Systems Center.)*

Toward a National Intermodal System

A recent report from the National Commission on Intermodal Transportation, directed by Anne Aylward, describes the advantages to the United States of being aggressive in the intermodal world and issues that must be dealt with for this to take place [1].

The report emphasized the international aspects of intermodalism and the fragmented policy structure within the United States DOT for dealing with intermodal issues. DOT has modal administrations: the Federal Highway Administration, the Federal Aviation Administration, the Federal Maritime Administration, the Federal Railroad Administration, and so on. The cross-cutting questions of intermodal investment tend to be de-emphasized.

This report stressed the fact that institutional change is hard, but that true intermodalism will require it. It considered whether technology expedites or inhibits change in this context, and also considered the Congressional constraints in the form of individual congressional committees which guard their prerogatives fiercely—in addition to the DOT constraints mentioned earlier—that can inhibit intermodal development.

Air Freight

Air freight is another mode that is an increasingly viable option. Air freight is very costly. It charges high prices for very high-quality services such as "just-in-time" emergency deliveries. While it has a very modest overall share of the market, it is growing very quickly. This is expensive transportation—while less than 1% of the ton-miles in the United States are carried by air, air freight has more than 4% of United States freight revenues. Both UPS and Federal Express have been pioneers in the air freight business providing dedicated service for certain kinds of shipments, moving containers by truck to and from airports for intermodal service.

Cost/LOS Trade-Offs for Various Modes

The level-of-service versus cost trade-offs for the various modes are shown in Figure 20.7.

Figure 20.7 *LOS versus cost for various freight modes.*

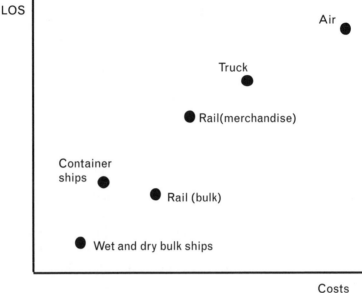

Figure 20.7 *LOS versus cost for various freight modes.*

These services can be put together to form intermodal moves. A key issue, we should note, in intermodalism is how to split the revenues—those monies received from the shipper and receiver—among the intermodal partners. As you would imagine, this is usually a hard negotiation.

Another related question is who deals directly with the customer. In truck/rail intermodalism in the United States, the truckers have developed the relationship with the customer at a retail level; the railroads provide wholesale service to the truckers. It is fair to say the railroads are concerned that the truckers have the direct customer relationship. What they can do about it is not clear.

Freight Summary

We have discussed different freight modes, technology advantages of each, different kinds of markets—bulk commodities, merchandise, etc. We discussed intermodal initiatives in current-day freight transportation. Internationalism is a critical and quantitatively different element in the transportation world compared to twenty years ago.

Key Factors

The key factors in freight transportation modes are as follows:

- The cost structure: relationship of fixed and variable costs;

- The nature and ownership structure of the physical assets: infrastructure, terminals, vehicles;

- Technology;

- The regulatory framework;

- The structure of the market.

There are big structural differences in both ownership and technology in the modes. When thinking about these various modes, consider the regulatory framework within which these modes operate. How easy is it to enter and exit various services? How regulated is pricing?

Vehicle-Cycle

In all modes, recall the vehicle-cycle as a fundamental element of productivity.

Vehicles and Infrastructure

We recognize the importance of understanding the interrelationship between vehicles and infrastructure. For example, if one discusses double-stack container trains, which improves productivity substantially by reducing fuel and crew costs per container, we need also to consider the infrastructure needs—bridge and tunnel clearances and wear and tear on the rights-of-way due to heavier loads. Heavily weighted trucks may have a disproportionate impact on highways, raising maintenance costs. In the trucking industry, longer combination vehicles (LCV), where trailers travel in tandem behind a single tractor unit, have an efficiency versus safety trade-off.

The Market

We need to understand the characteristics of the marketplace. What are the level-of-service variables that the customer is interested in? What is the competitive structure of this industry? What is the network structure? What is the location of economic activity that generates traffic?

Operating Plans and Strategic Plans

What kinds of operating plans do we need to operate effectively? What are the fundamental trade-offs—cost versus LOS—in the way in which we operate; what does the concept of plan integrity mean in different modes? How much does flexibility in the operating plan cost? How is the strategic plan (capital assets) linked to the operating plan?

Intermodal, international, global transportation companies—looking at new kinds of customers and relationships which require customized services—need flexibility in their operations. This flexibility is provided through technology (information systems in particular) and through management.

The 30 Key Points

Think about the different modes and intermodal competition and how they relate to our 30 key points. A good exercise would be to study those key points and think about how they relate to different freight modes.

Think about the triplet: technology, systems, and institutions, and how we deal on these three dimensions to achieve competitive modern freight transportation services.

We now proceed to the section on traveler transportation.

REFERENCE

1. National Commission on Intermodal Transportation, "Towards a National Inter-Modal Transportation System," Washington, D.C., September 1994.

PART III

Traveler Transportation

CHAPTER 21

Traveler Transportation: Introduction

Traveler Transportation

We will discuss many issues, and both conceptual and modeling questions associated with traveler transportation. But, as we will see, there are a lot of fundamental characteristics of transportation systems that are common to traveler and freight. For example, the fundamental stochasticity of demands and operations in traveler transportation is as important as it is in freight systems. The notion of network structure and how links and nodes fit together cuts across freight and traveler systems. When we talk about level-of-service, many of the parameters for traveler systems are similar, if not identical, to those in freight systems.

Differences Between Traveler and Freight Transportation

Nonetheless, there are some important differences between traveler and freight; we begin by considering some of these differences.

The Transportation Process

First, the process of going from origin to destination is much more important for travelers than it is for freight. The cargo is people—they can smell the environment, hear loud noises, get upset if the lights on the subway car go out for a long period of time—as opposed to inanimate freight that is not sensitive to those issues. Freight customers are primarily concerned with timely and reliable delivery in a state ready for a subsequent part of the manufacturing process or the sales process.

The process for a traveler is important. How one measures service quality going from Boston to Washington is not governed simply by how long it takes. The traveler is concerned with comfort, with politeness of the ground and air personnel, with how long one sits on the runway, and so forth.

Safety and Security

Second, safety and security are typically much more important for travelers than for freight. So safety and security are areas that traveler operators are typically willing to spend more money on than freight operators. Human life is highly valued—we say it is "priceless"; this implies that expenditures for protecting it are much greater than for protecting even quite valuable cargo.

Now, having said that human life is "priceless,"— and indeed in an ethical sense it is— in fact, there is a price on it. And as harsh as that is to say, it is true that one can look at investments and expenditures that operators and regulators insist upon for safety protection in transportation, and one can compute how much money is being spent to save a human life. Implicitly, we are willing to spend a certain amount to save a life; if it costs more than that amount to save it, we do not. While the press and public does not like to hear that, it is central to safety-related decision-making in transportation systems and other systems as well—environmental protection, for example.

Level-of-Service Variables

Third, there are differences in traveler and freight level-of-service variables. When one analyzes freight level-of-service, we use an economic basis for deriving the important level-of-service variables. We use the inventory concept and logistics costs as a mechanism for understanding why average travel time and the variation in travel time are important in the selection of a mode by a shipper. The probability of loss and damage for a shipment can be factored into a decision about whether to use a particular mode based on economic—or, if you will, business—considerations.

In traveler transportation, often the level-of-service variables can be more subtle and psychological in nature. Of course, they can be quite quantitative with economically-based level-of-service variables. But, often, it is very difficult to capture issues that are important to people in economic terms, as they make judgments about the modes to use.

People seem to make judgments that are very hard to explain on an economic basis. The classic example is auto ownership; people purchase automobiles, where even the roughest economic computation would suggest that it does not really make any sense for this particular individual to own a car. There may be modes that are available that are cheap and reliable, and the person may have other pressing nontransportation needs, and yet the car is purchased. In that particular instance of car ownership, transportation choice includes psychological factors like self-image and why it is so important—if your neighbor has a car—for you to have a car. This may lead to decisions that, on a narrow economic basis, would be difficult to explain.

Now, we introduced the notion of utilities earlier; we discussed comfort in transportation systems and we used the variable called "hugs" as a way of dealing with issues that are very difficult to quantify but nonetheless are real. In traveler transportation people are making judgments based on some psychological factors that have no equivalent in freight decision-making.

Groups

Fourth, humans exist in groups—families, fraternities or sororities, or groups of individuals who live together and may make transportation decisions collectively; often these decisions are made in fairly complex and subtle ways. How many cars should a suburban family with teenage children have? This is a complex and difficult-to-quantify decision.

Motivation for Travel

Fifth, travelers typically have many different reasons for traveling. We travel to get to MIT, to go down to Cape Cod on vacation, to California to see our parents, and so forth.

In shipping freight, the reason for the trip is the same every time. When the coal mine operator ships coal to the power plant, she does it for the same reason every time: the coal must be transported from the mine to the power plant, so electric power can be generated.

However, John Smith makes trips for a number of different purposes. He goes to work; he goes shopping; he picks up his children at day care, and so on. Particularly in recent years, with so much concern about personal time constraints, people link trips together—"trip-chaining." The work-to-home trip combined with pick-up-children-at-day-care trip, combined with a trip to McDonald's, is an example. So the trip may be a complex chain for various purposes. Further, mode choice depends on trip purpose.

Travel as Discretionary

Sixth, for travelers, transportation can often be viewed as a discretionary item. Indeed, transportation may compete with other uses of the family budget. We have limits on money to spend: we will spend some on transportation; we will spend some on education; we will spend some of it on food, and so on. A trip to the movies is discretionary. We can make a judgment not to make that particular trip.

In freight, transportation is typically quite fundamental to the purpose of the business. Transportation of that coal from the mine to the power plant is fundamental to that business. It is not discretionary. If we decide that we will not transport the coal, we are not in that business anymore. So, the substitutability of nontransport for transport uses of resources that exists in traveler transportation often does not exist in freight, at least in the short run.

Success in the Marketplace

Seventh, a corollary of the previous point, is that the people who are concerned with shipping freight often have their success in the marketplace intimately tied to effective transportation choices. One of the fundamental decision variables in some businesses is what transportation mode to use: we make the tactical judgment to use the air freight system to keep the assembly line running today; we make the strategic judgment to switch from truck to rail because rail is cheaper and the quality of service is adequate. There are competitive forces operating in business so that bad transportation choices may eventually force companies out of business. Companies who are ineffectual in making transportation judgments will be less competitive. So the less competent organizations get pushed out of the business.

In a family group this phenomena is much less evident. Inefficient families are not squeezed out by the marketplace. If parents are not chaining their trips together intelligently and are making extra trips to the day-care center or to pick up the children after basketball practice, they may be spending more money on transportation than they ought; however the marketplace is not going to squeeze them out of business; we are not going to fire the parents because they are not very good at trip-chaining. So traveler transportation is a more chaotic market. There is no rationalization that takes place because inefficient firms are leaving the marketplace. Family groups or fraternity brothers have a lot of other ways of judging the quality of people within the group that are much more important than being good at trip planning.

Substitutability of Communications and Transportation

Eighth, from a technological point of view, the substitutability of communications for transportation is an important trend. The idea is that instead of traveling one can communicate; data can flow rather than people. This technological change is more germane in traveler transportation than in freight. It will have some impact on freight. We may be able to do a better job of monitoring freight on the network. But, ultimately, the coal has to be physically transported from the mine to the power plant.

However, people telecommuting, working in their home and coming into the office once a week rather than five times a week, is of increasing importance. Now, how this will actually affect demand for traveler service is an important question. The two opposing perspectives on it are:

1. Communications will greatly reduce the need for transportation because of the telecommuting option; people will not have to actually physically be at the office to make a contribution. With fax and e-mail capabilities at home, one can work as well there as in the office—maybe even a little better if your home environment is nice and quiet.

2. On the other hand, while telecommuting may occur, the additional economic interactions that will occur as a result of enhanced communication may generate more travel than is saved by the telecommuting option.

Transportation/communications substitutability is a research question. Research on telecommuting—understanding what kinds of firms adopt it and what kinds of people use it—is ongoing.

An important question: what are the implications of telecommunication advances for the level of transportation infrastructure investment?

Are there other differences that you might identify that I did not?

Isn't pricing of services very different in freight and traveler transportation?

That is a good point. In both freight and traveler there is flexibility in pricing. In freight operations there are often long-term contracts between shippers and carriers. The shipper gets a long-term fixed price; the carrier gets guaranteed loads. You see that less in the passenger business, although one can buy a monthly transit pass.

What about differences in the ability to choose modes? Isn't it more difficult in freight?

Modal choices are probably more limited in freight transportation than they are in traveler transportation. If you are that coal mine operator, you do not have many transport choices. What you want to do is get the lowest possible price from the railroad—loading your coal into trucks does not make sense; on the other hand, a traveler can make choices among public transportation, automobile, and so forth.

Aren't level-of-service expectations very different?

The scale of level-of-service variables is quite different; service expectations for travelers are much higher than they are even for the most high-value intermodal freight. If you commute by commuter rail, a 15-minute delay is important. There is not much freight where a 15-minute delay is a major problem.

Equity issues are more important for travelers than freight.

True—if a group—the elderly, say—does not have access to transportation, that gets the public's attention. Freight shippers have less political visibility, but they band together to advocate for better service.

> Passengers can provide their own intelligent movements, going from a bus to a train, say. Freight going from one mode to another has no intelligence.

This is correct. On the other hand, we can electronically track freight; doing that with people creates privacy problems.

....

Traveler Transportation Statistics

Next, we present some statistical information on the United States traveler transportation system.[1]

The average number of traveler-miles per year per person in the United States is about 15,500. 15,500 miles a year is 3.8 trillion traveler-miles for the United States.

Mode Choice: The Dominance of Cars in the United States

Traveler transportation in the United States reflects the fact that it is an automobile-oriented society; on the order of 90% of the traveler-miles are highway and the overwhelming share of that is private automobiles.

The other modes: local transit, intercity rail, airlines, and water-borne traveler services together make up the remaining 10%. The largest fraction of that is in airline transportation, with about .3 trillion passenger miles; local transit has .04 trillion miles; the others are very small (see Figure 21.1).

What is an example of water traveler transportation?

> Commuter boats.

Yes—there are services in a number of cities—Seattle, Washington, for example, taking people who live on the harbor islands downtown. In the Boston metropolitan area, a commuter boat from the South Shore coastal community of Hingham, Massachusetts, goes across Boston Harbor to downtown. The highway commute from the south to Boston is particularly difficult with congested roads—so the use of

---

1   Statistical data is taken from *Transportation Statistics Annual Report 1994*, Bureau of Transportation Statistics, U.S. Department of Transportation, Washington, DC, January 1994.

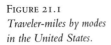

FIGURE 21.1
*Traveler-miles by modes
in the United States.*

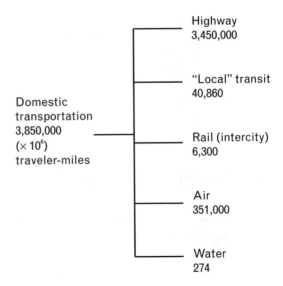

Domestic
transportation
3,850,000
($\times 10^6$)
traveler-miles

Highway
3,450,000

"Local" transit
40,860

Rail (intercity)
6,300

Air
351,000

Water
274

water-borne transportation in the Boston metropolitan area is a good option for some commuters. In the developing world, in cities like Bangkok, Thailand, boats on "klongs" (canals) are a vital mode.

Land Use and Highways

Why the dominance of highways in traveler transportation? It is connected fundamentally to the kind of land use that has evolved in the United States. In short, the automobile enabled the development of spread-out land-use patterns. These patterns can only be served by automobiles in many cases; low densities make public transportation less than viable. Environmental degradation may result from the dominant use of automobiles, especially with respect to air quality.

Suburbanization

Two major trends have taken place over the twentieth century. First, there was a massive move from rural areas into the cities; people moved from the farms to manufacturing jobs in urban areas.

Second, the United States has become a suburban nation; people moved from the core city to suburbs on the periphery of metropolitan areas. In 1950 about 23% of the population of the United States lived in the suburbs. In 1990 the comparable number was about 46%. The idea of people living in suburban settings is powerful. They presumably get a mix of amenities and practicalities. On the one hand, they have the urban advantages of living close to a large population center with the jobs and cultural attractions that it has; on the other hand, they get a rural perspective—the notion that "everybody" wants to be able to look out on their own patch of lawn, see the sky, breathe fresh air, and have peace and quiet.

The idea of not living in the core city but having a plot of land with some privacy is not a new idea. Max Lay, in his definitive history of land transportation, describes such ideas in antiquity [1]. The growth of the suburbs is linked to the auto-oriented kind of society that we have in the United States. This "sprawl" is not without cost, such as transportation energy use, congestion, air-quality issues and, often, the poor left behind in the urban core with no cars and no access to jobs.

In the United States tax code, interest on home mortgages is tax-deductible. So we can talk about people wanting to breathe fresh air, but there are other arguments based on economics. The interest on a home mortgage is a tax deduction. The interest on an educational loan is not. It is believed that this is good public policy—it gives an incentive to invest in a home and it generates construction jobs. It also generates a spread-out land-use pattern that is very difficult to sustain from an energy and environmental point of view, according to some [2].

So, growth in core cities, with people attracted from the hinterland, occurred as the United States changed from an agrarian nation with farming being the way in which a large fraction of the population supported themselves, into a manufacturing and services economy. Then, we saw development at the urban periphery. What has happened over a longer period of time is that the core city has spread out, both politically and economically, to include the inner suburbs. This led to another round of suburban development still farther out from the core as we spread our land-use over a still larger area.

A Brief History of Metropolitan Areas

Consider the history of some of the metropolitan areas in the United States, particularly in the northeast, where cities are older by United States standards. For example, consider an inner suburb like Revere, Massachusetts, north of Boston. At the turn of the century, Revere was a playground for wealthy people. It is a beachfront community. People with money summered in Revere in much the same way they summer in a Maine vacation spot today. Now Revere has been absorbed into the broader Boston metropolitan area. Frankly, it is not viewed as a vacation spot for hardly anybody at all; it is a relatively economically depressed area.

Coney Island in Brooklyn, New York, at the turn of the century was a playground for people in New York City. It is in the southern part of New York City, perhaps 20 miles from Times Square in midtown. In 1880 Coney Island had three major racetracks for thoroughbred horses, all long gone. And Coney Island is now certainly an integral part of New York City.

Consider the Harlem area of Manhattan Island, which is at about 125th Street in the northern part of Manhattan. 100 years ago that was a suburban area. The population of Manhattan was primarily south of 42nd Street, and Harlem, which has had its social woes, and certainly is an integral part of New York City, was a place a century or more ago where people went to "get away from it all."

Commuter Lines and Garden Cities

Max Lay insightfully notes that the initial development of the inner suburbs was not automobile-driven, but rather was originally driven by the development of fixed-rail commuter lines out to peripheral areas around the city. Lay talks about the rail spokes coming out of the city enabling this spread-out land-use pattern (see Figure 21.2). The idea was that you would have—to use Lay's term—"garden cities" around the periphery of the core city; these garden cities were to be designed at "human scale", you could walk around them; they were self-contained; they were designed to allow amenities that one could not achieve within the core city.

The building of these rail lines was financed by land speculators depending on increased real estate values as transportation accessibility increased. As land becomes more accessible, as a result of a new rail line, the value of that real estate will go up. Fares on these lines were kept low to assure that garden cities would develop. But these fares could not cover transportation operating costs and, as Lay states, "the car was ready to take up the slack and sustain the suburban sprawl promoted by land speculators."

FIGURE 21.2   *Core and garden cities.*

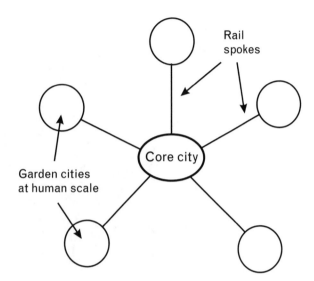

Eventually, "infill" took place between the rail spokes because of the accessibility that automobiles could provide through the flexibility and nimbleness of that technology.

Thus, urban sprawl developed. The urban sprawl came about not because of the original development of the rail lines, but rather because of the subsequent development of automobile technology, which allowed one to fill in the spokes. Sprawl, which was not of human scale in the way that garden cities with walkable distances (what today would be called transit-oriented development) were intended to be, was the result. And now the trend toward new development further out on the periphery only exacerbates sprawl (see Figure 21.3).

On Long Island, New York, there is a city named Garden City. It is accessible by the Long Island Railroad, which is a transportation mode that serves thousands of people every day going from the Long Island suburbs into New York City. These so-called garden cities became less viable because they were no longer walkable, self-contained suburbs with a high quality of life. They were beyond human scale. They were overwhelmed by the automobile's impact on a regional scale.

And, at the same time, over the years there have been substantial developments of inner-city urban expressways, the Interstate system within the urban core, which Lay argues is also beyond human scale.

FIGURE 21.3   *"Infill" between the "spokes."*

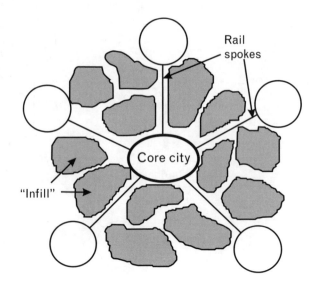

Mega-Cities

The trend worth noting is the growth of "mega-cities," particularly in developing countries. These gigantic cities grow as societies become less agrarian and more focused on manufacturing. People gravitate to the city to create the labor pools necessary for these manufacturing operations. A recent article in *The Economist* pointed out that in 1950 there were only two cities with populations greater than 10 million people: London and New York. In 1994 there were 14 such cities. They project that in 2015 there will be 27 such cities, including many from the developing world, such as Shanghai, Bangkok, and Lima.

Ring-Roads

There are other interesting issues relative to land use and the way in which urban form develops. For example, we have so-called "ring-roads" that ring cities—Boston with Route 128, and many other cities around the United States with ring-roads servicing transportation needs on the periphery. This has negative implications on our ability to provide public transportation services, either through bus services or rail services, since low-density land use does not create origin-destination volumes needed to sustain public transportation. The spread-out land-use forms make transportation by modes other than the automobile very difficult.

Edge City

An excellent book called *Edge City* by Joel Garreau [3] suggests that every United States city that is growing at all is growing in the fashion of Los Angeles, with multiple urban cores. That is, we have the central core with varying degrees of health, depending upon the particular city; and then you have other cities that are developing around that urban core. In the Boston area, Framingham, Massachusetts, is an "edge city." He talks about major shopping malls that have become so popular in the last several decades, as the equivalent of the "village squares" in the garden city concept. These malls, he argues, are the equivalent in today's society to village squares, where everybody comes to meet and intermingle. For example, many elderly citizens use the malls, before they open for business, as walking tracks out of the weather.

Garreau discusses moving of residences from the core city to the suburbs—people deciding they want to live there—followed by stores moving out to the suburbs. If that is where the people live, that is where Filene's wants to locate—not downtown, but out at the Burlington Mall. Finally jobs move out to the suburbs, leading to the core city itself becoming substantially underdeveloped and underutilized, with the economic action taking place at the periphery and poor people trapped in the core city with no way to conveniently access the jobs in the suburbs.

Garreau is very concerned about the loss of the "sense of place", people do not have a feeling of where they are; everything is the same; all areas have become homogenous, to the detriment of all of us, he argues.

Land Use and Public Transportation

These land-use patterns and urban forms are a concern of the Federal Transit Administration (FTA) of the U.S. Department of Transportation, which is responsible for public transportation at the federal level, and questions whether it will ever be viable to develop transit systems in the face of these kinds of land-use patterns.

The fundamental idea is that you cannot separate transportation policy from the way in which land is used: for residences, for shopping, for jobs. Land use and transportation are hand-in-glove.

To quote from a recent study on the perspective of the Federal Transit Administration: "The United States has undergone 50 years of rapid urban growth and suburban development that have radically transformed the landscape and the environment. Most people live in or on the periphery of metropolitan areas, many of which have grown in land area far faster than population. Suburban 'sprawl' and its by-products of traffic congestion, air pollution, loss of open space and other environmental degradation, are familiar issues in all parts of the country, as are concerns for decaying inner cities, municipal competition for tax revenue, and sharp disparities in economic opportunities and public health. While many people unquestionably are enjoying substantial benefits from the current patterns of metropolitan land use, others are worrying about the consequences of continued current development trends." This last sentence says there are both benefits and costs in sprawl. As always, who benefits and who bears the costs is a key question.

The report goes on to say that "many of the decisions that shape metropolitan land-use patterns lie in the hands of individual land owners,"—this is a democracy. People, to a first approximation, can do with their land as they see fit—"developers, businesses, and local government, the federal government, and Congress have substantial interest in the results and the Federal Transit Administration (FTA) is responsible for assuring that tax dollars spent on mass transit systems yield high social and economic return on taxpayer investment." The report then suggests that the low-density development patterns make providing public transportation services extremely difficult.

Multidisciplinary Approach

When the FTA discussed the studies they anticipate performing, they referred to an interdisciplinary set of professionals that might participate in studying this kind of a problem. They say here that experts will come from such fields as urban policy, real estate development, regional economics, municipal finance, landscape ecology, transportation, urban air quality, public health, and civil engineering. They cite the need for experts drawn from many disciplines; you will find, as you proceed professionally, that more often than not you will be working with people in areas of expertise very different from yours.

The New Transportation Professional

What will it require to be a transportation professional in the future—on into the new century? There is a lot of debate about and differences of opinion about that, as you might expect, nationally, and internationally. But what everyone agrees on is that being a transportation professional will be more complex and more interdisciplinary in nature, with new technologies and with new interactions between the transportation system and various socio-political-economic issues. Four decades ago, being a transportation professional meant being able design a physical facility that would support transportation activity, like a highway or a railroad; we have seen an enormous evolution from that, as people now consider fields like operations research, systems analysis, modern technology, and institutional and management issues as fundamental. The idea of the "T-shaped" transportation professional, combining breadth and depth, has been advanced by Sussman as a framework for the future of transportation education [4] (see Figure 21.4).

REFERENCES

1. Lay, M. G., *Ways of the World: A History of the World's Roads and of the Vehicles that Used Them*. New Brunswick, NJ: Rutgers University Press, 1992.

2. *Brookings Review*, Special Issue on "The New Metropolitan Agenda," Brookings Institution Press, Fall 1998.

3. Garreau, J., *Edge City: Life on the New Frontier*, Doubleday, 1991.

4. Sussman, J. M., "Educating the 'New Transportation Professional.'" *ITS Quarterly*, Summer 1995.

FIGURE 21.4
The T-shaped "new transportation professional."

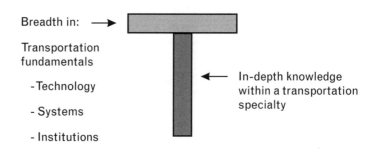

Breadth in: →

Transportation fundamentals

- Technology

- Systems

- Institutions

← In-depth knowledge within a transportation specialty

CHAPTER 22

Commuting, Nonwork Travel and Safety, and Some Transportation History

....
Commuting

We now turn to commuting. Traveling to and from work is an important trip purpose. We have seen some real changes in the pressures on transportation systems by people going to and from the workplace. There has been an expansion from 1980 to 1990 of about 10% in the number of people that go to work—on the order of 19 million additional workers in 1990 compared to 1980. A major demographic force is women in the workplace.

How does commuting take place in the United States?[1] Figure 22.1 gives a breakdown.

Suburb-to-Suburb Commutes

The largest flows are in "suburb to suburb," 32 million trips and the commutation into the central city of about 43 million (when you add 16 million from the suburbs to 27 million originating and terminating in the central city). You get a sense of the importance of suburbs in the economic activity in the United States.

Over the last 10 years, as the number of workers and the number of work trips have grown, there are some interesting trends. First of all, the percentage of people who are driving alone, the "single occupancy

---

1 Pisarski, Alan E., "Commuting in America II," ENO Transportation Foundation, 1996, is the definitive reference on this topic.

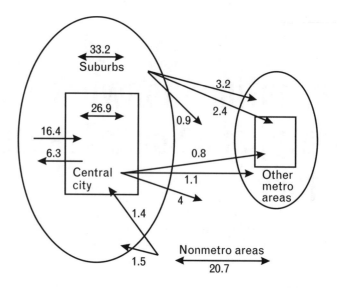

FIGURE 22.1   *National commuting flow patterns (millions of daily trips) (Source: Transportation Statistics Annual Report 1994, Bureau of Transportation Statistics, United States Department of Transportation, Washington, DC, 1994.)*

vehicle" (SOV), has grown from about 60% of journey-to-work travel to about 70%. The number of people in car pools over the last ten years has dropped from about 20% down to about 15% of work trips. So the drive-alone mode has become more important. People like the flexibility in SOV, especially with the lifestyles many people have requiring trip-chaining and with current inexpensive energy (gas).

Public Transportation

The public transportation share of work trips nationally has shrunk from a modest 8% in 1980 to an even more modest 6.5% in 1990. Driving alone is becoming more popular and car pooling is becoming less popular; in fact, everything is becoming less popular except driving alone, as people react to what they see as the convenience and maybe even the cost advantages of single occupancy vehicles, given today's land use patterns.

Nonwork Travel

Consider nonwork travel. Over the last several decades, the growth rate of nonwork travel by travelers, by drivers, transit users, etc., has exhibited a faster growth than work travel, which itself exhibited fast growth. The demographics of the American family are changing, with household size declining, the number of households increasing, women in the workplace, two-job and multi-job households. All of these lead to substantially more travel of both a work and a nonwork nature. We have mentioned trip-chaining—people putting together various kinds of trips, work trips and nonwork trips, in complicated ways that allow them to efficiently route themselves around the system, taking care of many purposes in one chained trip.

Table 22.1 contains some interesting statistics for the United States:

TABLE 22.1   Average Daily Trips, Travel per Person and Person Trip Length (by Sex and Trip Purpose) (*Source:* Transportation Statistics Annual Report 1994, Bureau of Transportation Statistics, United States Department of Transportation, Washington, DC, 1994.)

	MALE	FEMALE	TOTAL
AVERAGE DAILY PERSON TRIPS			
Earning a living	0.77	0.57	0.66
Family and personal business	1.12	1.42	1.28
Civic, educational, and religious	0.34	0.36	0.35
Social and recreational	0.78	0.74	0.76
Other	0.02	0.02	0.02
Total	3.03	3.12	3.08
AVERAGE DAILY PERSON MILES OF TRAVEL			
Earning a living	10.5	5.12	7.69
Family and personal business	8.62	9.23	8.93
Civic, educational, and religious	1.79	1.89	1.84
Social and recreational	10.4	9.38	9.86
Other	0.25	0.20	0.22
Total	31.56	25.83	28.56
AVERAGE PERSON TRIP LENGTH			
Earning a living	13.91	9.18	11.8
Family and personal business	7.75	6.63	7.11
Civic, educational, and religious	5.39	5.38	5.38
Social and recreational	13.45	12.93	13.19
Other	11.59	9.13	10.3
Total	10.54	8.47	9.45

The average number of trips taken per person in the United States for 1992 is on the order of three per day. Three is the national average and differs very modestly between male and female (for male—3.03, for female—3.12). Even with the growth in women in the workplace, though, the numbers of trips taken for business by men are on the order of 20% more than those taken by women.

The overall average miles of travel is about 28.5 miles per day. Again, although the number of trips does not differ very much between male and female travelers, the trip lengths do. The average male travels 31.5 miles a day. The average female travels 25.8 miles a day. The distances to travel for various purposes vary with gender as well. The average work trip in the United States is about 11.8 miles, one way. That differs, interestingly, substantially between women and men again. The average man is traveling 13.9 miles to work; the average woman 9.1 miles.

Intercity Travel

We now consider intercity travel (see Figures 22.2, 22.3, 22.4). Business and personal trips differ importantly. On the order of 16% of the trips of over 100 miles are business-related; the remainder are

FIGURE 22.2 *Intercity travel by purpose: 1991. (Source: Transportation Statistics Annual Report 1994, Bureau of Transportation Statistics, United States Department of Transportation, Washington, DC, 1994).*

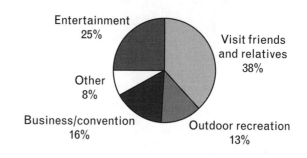

FIGURE 22.3 *Modal choice by trip purpose: 1991 (Source: Transportation Statistics Annual Report 1994, Bureau of Transportation Statistics, United States Department of Transportation, Washington, DC, 1994).*

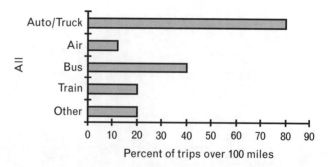

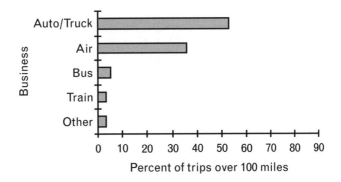

FIGURE 22.4   *Air and auto trips by trip length (Source: Transportation Statistics Annual Report 1994, Bureau of Transportation Statistics, United States Department of Transportation, Washington, DC, 1994).*

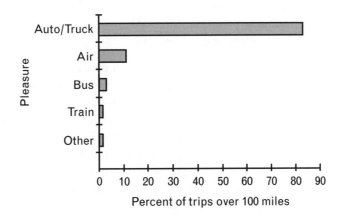

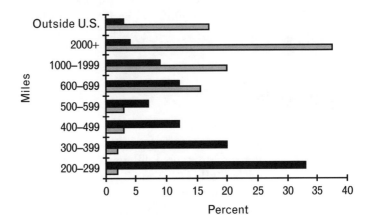

personal trips—entertainment, outdoor recreation, visits to family and friends, and other purposes. The average length of business travel, above that 100-mile cutoff, is 862 miles. The average personal trip, above 100 miles, is about 800 miles.

Again, the dominant mode is overwhelmingly automobile, even for longer distance. If one looks at all business trips above 100 miles, the number taken by automobile is over 50%. The number taken by air is about 35%. So even on the business side, auto is strong. If one looks at all trips, including pleasure trips, 80% are by automobile, with only about 15% air. So it is still primarily an automotive world; one has to have rather lengthy trips before air starts to dominate.

International Travel

However, in international travel, the air mode dominates. The United States is the major international destination in the world. There are about 48 million annual visitors to the United States. About 20 million of those 48 million are non-North American—that is, other than Canada and Mexico, which are bordering countries. About 17 million are from Canada. About 10 million are from Mexico.

Forty-eight million is about 28% of the population of the United States. There are any number of European countries that generate well over 100% of their population each year in international visitors. So, while the United States has a high absolute number—48 million is the largest in the world—as compared with population, it is more modest than it is in countries like France and England and others in Western Europe.

Safety by Mode

Next, consider safety, an important level-of-service variable. Certainly there are substantial differences in the safety of the modes that travelers use (see Table 22.2 and Table 22.3).

Automobiles are substantially less safe than other modes by various measures. There has been a consistent decrease, though, in fatality rates in automobiles in the U.S., over the last decade or so—from 3.2 fatalities per 100 million vehicle-miles in 1981, to

TABLE 22.2   Trends in Fatality Rates by Transportation Mode: 1981-1992 (*Source: Transportation Statistics Annual Report 1994*, Bureau of Transportation Statistics, United States Department of Transportation, Washington, DC, 1994.)

Year	United States Air Carrier (per million miles flown)	General Aviation (per 100,000 hours flown)	Motor Vehicle (per 100 million vehicle-miles)	Railroad (per 100,000 train-miles)	Recreational Boating (per 100,000 boats)
1981	0.001	3.48	3.2	0.82	8.0
1982	0.008	4.01	2.8	0.89	7.6
1983	0.005	3.72	2.6	0.89	7.9
1984	0.001	3.58	2.6	1.01	6.5
1985	0.054	3.37	2.5	0.80	6.7
1986	0.001	3.56	2.5	0.84	6.2
1987	0.053	3.11	2.4	0.93	5.9
1988	0.063	2.91	2.3	0.84	5.1
1989	0.028	2.75	2.2	0.84	4.7
1990	0.008	2.67	2.1	0.98	4.4
1991	0.010	2.74	1.9	1.02	4.6
1992	0.007	2.98	1.8	0.99	4.0

1.8 per 100 million vehicle-miles in 1992—almost a halving of the fatality rate.

Now, of course, the change is not as substantial for the total numbers of fatalities, because of a continuing increase in vehicle-miles traveled. The fatalities on highways in the United States are hovering around 40,000 per year, but it is on a larger traffic base. If the fatality rate had not improved, we would have about 60,000 or 70,000 fatalities per year. This improvement is attributed to several causes: seat-belt usage, air-bags, tougher drunk-driving enforcement, and more crashworthy vehicles.

Within motor vehicles, though, there are extraordinary differences in safety for different vehicle types. For passenger cars, the numbers of deaths per 100 million vehicle-miles is about 1.3. For trucks, it is about 1.5, about 12% higher. For motorcycles, it is 25.1 deaths per 100 million vehicle-miles (!)—on the order of 20 times the fatality rate for automobiles!

TABLE 22.3   Motor Vehicle Fatalities by Major Vehicle Type: 1981–1992 (*Source: Transportation Statistics Annual Report 1994,* Bureau of Transportation Statistics, United States Department of Transportation, Washington, DC, 1994.

YEAR	PASSENGER CARS		TRUCK		MOTORCYCLES		PEDAL-CYCLISTS	TOTAL NON-OCCUPANT
	Deaths	Rate*	Deaths	Rate*	Deaths	Rate*	Deaths	Deaths
1981	26,645	2.4	7,343	1.8	4,906	45.9	936	8,877
1982	23,330	2.0	7,303	1.8	4,453	44.9	883	8,299
1983	22,891	1.9	7,184	1.6	4,265	48.7	839	7,746
1984	23,621	1.9	7,570	1.6	4,608	52.5	849	7,973
1985	23,214	1.8	7,666	1.5	4,564	50.2	890	7,782
1986	24,944	1.9	8,243	1.6	4,566	48.6	941	7,853
1987	25,132	1.9	8,910	1.6	4,036	41.0	948	7,825
1988	25,808	1.8	9,217	1.6	3,662	36.5	911	7,917
1989	25,046	1.7	9,402	1.6	3,243	30.1	832	7,495
1990	24,092	1.6	9,306	1.5	3,244	33.9	859	7,465
1991	22,385	1.4	9,052	1.5	2,806	30.6	843	6,768
1992	21,366	1.3	8,666	1.4	2,394	25.1	722	6,366

*Death rate per 100 million vehicle-miles of travel

Some Transportation History

Max Lay's book contains interesting historical discussions of the evolution of transportation.[2] Lay begins with carts, carriages, and stage coaches, and the use of horses for transportation. The environmental impact, particularly in urban areas, of horse-drawn carriages, was substantial—due not only to the waste matter left behind by the horses, but to the fact that these horses had a tendency to die under the stress and strain of their efforts. The physical removal of the carcasses of these dead horses was also a major environmental and public health issue of the day.

2   Lay, M. G., *Ways of the World: A History of the World's Roads and of the Vehicles that Used Them,* New Brunswick, NJ: Rutgers University Press, 1992. This is an invaluable reference for those interested in more detail on the history of surface transportation.

So, the environmental impact of transportation is hardly a new issue. Looking at it from a systems point of view, the power is biological in nature—a horse; the energy is provided by the horse's food; there is waste as well. The energy produced by the horse is providing for your transportation.

Transportation was extremely expensive. As Lay mentions, the cost of a round-trip from Paddington Station to another station in London in 1800 by horse-drawn vehicle was about 1% of the annual wage of a typical worker, where the comparable trip today would be about .02% of the annual wage of that same worker. Indeed, dominant trends in traveler (as well as freight) transportation over a long period of time is the substantial lowering of costs.

System safety was several orders of magnitude poorer in earlier times than it is now. On United States highways, 40,000 people a year die. But if one looks at the safety of transportation systems in these earlier times, they were less safe than today, when adjusted for distances traveled.

Why do we travel, if it is so unsafe? Of course, we have talked about the societal and economic reasons for transportation. An anecdote somebody once told me stuck in my mind, relative to the question of why we travel if it is so unsafe. If we traveled less because good safe infrastructure was not provided, we would have discovered penicillin much later. Building a scientific enterprise would have been more difficult; we would have had antibiotics later. From an overall systemic point of view, human society would have been much more unsafe for the time interval before penicillin compared to those few hundred thousand people a year that were killed in transportation accidents. So one has to look at it more systemically.

What Enabled Transportation to Advance?

What pushed transportation forward? Drawing further on Lay, let us consider several important factors.

Technological Developments

A major revolution in urban traveler transportation—beyond the horse—was the development of electricity—powering trolleys and other vehicles running on rails on urban streets. Electricity allowed much faster speeds and higher reliability at much lower costs. It was the introduction of electricity, and trolleys running on fixed rails in urban areas, that changed urban geography with growth along corridors emanating out from the city center.

In the latter part of the nineteenth century, bicycles were invented and developed, based on advances in material and manufacturing techniques. These were an important mode in some United States cities and still are in some cities in the developing world.

Automobile Dominance

We have noted the relationship between land-use patterns—sprawl—and the dominance of the automobile. But there are technological and political roots as well.

The development of three technologies advanced the automobile to the forefront. Those were:

• The rubber tire;

• The internal combustion engine;

• The extraction of crude oil from below ground.

The Gas Tax

Politically speaking, a factor in the development of highway infrastructure is the relative ease of tax collection for this particular mode of transportation. People are taxed at the gas pump. People are taxed for ownership of vehicles. You pay, for example, in many states, an excise tax every year on your car, as a function of the value of a car. These taxes tend to be less objectionable than income taxes; collection is easy and comes in modest increments.

A question that many state secretaries of transportation are wondering about is what is going to happen if electric cars become prevalent, and people are not buying much gas? Where is the money to support the infrastructure going to come from, if we cannot collect at the gas pump, because people are buying electric power rather than gasoline?

Construction Jobs

Another political force behind highway development is that construction creates jobs. This is an enormous wellspring of employment opportunities in construction and related industries as well. For example, there is a large aggregate industry—sand and gravel—in this country.

The Transportation Industry

The transportation industry is a major lobbying force in the United States. Lay points out that for many years the largest companies in the United States were all directly or indirectly related to the automobile business. That has changed some with the advent of the information society, although we still have some very large transportation-related companies. Seven of the top ten companies in the 1950s, from a revenue point of view in the United States, were General Motors, Ford, Chrysler, Standard Oil of New Jersey, Mobil, Texaco, and Gulf—automobile and energy companies. There was, and still is, an enormous amount of political muscle.

The post-World War II period saw major highway investment and purchases of automobiles by people using money saved during the war, when there was little to buy (certainly no automobiles).

Finally, there was a broad consensus, up to the recent development of the environmental movement, that cars are good; the flexibility that they provide is good for the economy; it was the "American way." We were destined to drive around in single occupancy vehicles; it made a lot of sense to build highways to provide the job opportunities, social opportunities, and a variety of economic opportunities. So there was an enormous amount of momentum behind the development of highways.

Momentum behind highways has slowed in the past several decades. As noted above, the environmental movement was an important factor in that. Going back to the early 1960s, Rachel Carson, with her landmark *The Silent Spring*, sensitized millions of Americans to the environmental havoc that was being caused by various human activities, including transportation in automobiles.

Environmental Concerns

In the early 1970s, the "Stop the Highways" movement was in full cry, particularly in urban areas. Citizens in San Francisco stopped the building of the Embarcadero. Governor Sargent, in 1971 in Massachusetts, dictated that there would be no further highway construction inside Route 128. These two are examples of a new consensus that "building our way out of congestion problems" is no longer a viable political, economic or environmental option, particularly in our cities.

CHAPTER 23

Traveler Level-of-Service

Traveler Level-of-Service

In this chapter, we first discuss the dominant traveler mode, automobiles, and then generalize to LOS variables for travelers.

Why People Like Cars

Why do people like automobile travel? We take that as an article of faith. Certainly, this is true in the United States; in the developing world, it seems that as soon as per capita income rises, auto ownership jumps [1]. We now discuss why—from the point of view of the selection of a transportation mode and level-of-service variables—we like cars.

Clearly, we like the flexibility. You come and go when you want to. You are not dependent on the bus or the train. You want to come in to work a little early or leave a little late, you do so. The automobile network is universal. It defines our urban form. Trains go only where there are tracks. And buses go only on their routes. There is flexibility, both in time and in space, associated with the automobile mode. It often (but not always) is the fastest mode, depending on levels of congestion, time of day, and the available alternatives.

Travel time is a very important level-of-service variable for people (as it is with freight). However, computing the value of time, and how it may differ among individuals, and how it may differ even for the same individual for different trip purposes, is very subtle. In building analytic models of why people decide to use particular modes, travel time and the value of saving time is a key factor.

An interesting aspect of this has developed only over the last several years. One often associates being in a car with unproductive time: "all you are doing is driving and getting from here to there." But, in recent years, the communications revolution has allowed people to have telephones in their automobiles. It has allowed people to have fax machines and computers in their automobiles. So being productive while in a car may change the importance of travel time for some people.

Any further ideas about why we find automobile travel attractive?

Privacy—you have your vehicle to yourself.

Right. Exactly. Privacy is something that is cited quite often.

Automobiles suggest that you are at a higher level of society.

Good point. Certainly the notion of status associated with one's vehicle has become part of the ethic. You have heard the term "conspicuous consumption." People often consume very conspicuously on an automobile.

People simply enjoy the sensation of driving.

Yes, you are "at-one" with your car; person/machine interaction is important for some individuals. There have been some scholarly books written about the interaction of humans and their artifacts, their tools, their machines. Some people interact in a very visceral way with the car. Some people really like steering and driving a five-speed standard transmission, getting a sense of pleasure at their skill.

Several other points. First, as noted earlier, given the way our land-use patterns have developed, particularly in the United States, cars are virtually a necessity. There are areas where taking at least part of your trip without an automobile is virtually impossible. Land-use densities are so low that public transportation is not viable. The automobile is fundamental and a necessity of life, not a luxury, depending upon land-use choices that society makes.

One final reason we love cars is that it is very often a good transportation buy. It is a good value for your transportation dollar. You get this high-quality transportation service and it is a good buy, because somebody else is paying a lot of the costs for the infrastructure and cleaning up the environment. Your choice is economically rational. Highway transportation may well be cheap, or at least cheap relative to the level-of-service that is being provided to you, because of the way in which the costs of the highway infrastructure are paid for. This is a subsidized mode; so it is a good deal, as well as being a high-quality, convenient, flexible, private mode.

At the same time, people may not realize their true costs of automobile driving. Their travel decisions are often based on the assumption that the cost of the vehicle is a sunk cost—"I have to own the car anyway, and using it is a small incremental cost." It is the rare driver who includes the ownership cost per mile in a trip decision.

Traveler LOS Variables

Let us now generalize the discussion of cars to level-of-service variables for traveler transportation.

Average Trip Time, Reliability of Trip Time

Average trip time and the reliability of trip time are important in making a mode choice for travelers, just as for freight. It can often be much more subtle on the traveler side.

Value of Time

There have been learned treatises and a great deal of research done about the concept of so-called "value-of-time" and how one computes it when one designs a transportation system. Virtually all major transportation projects are justified on the basis of time savings. So how do we value this time savings?

If someone gets to work five minutes faster because of enhanced infrastructure, what is the economic value of that? How do we go about computing what five minutes of time is worth to the traveler and society? It is a very complex question. People have used various surrogates, such as hourly wage rates (typically multiplied by a constant less than one). How does one take the collective time that is saved on a transportation system used by travelers—small time savings multiplied by big traffic volumes—and estimate how that is valued by society (as well as by the individual)?

Aggregating Small Time Savings over Many People

In the inventory model for freight, we essentially mapped time into a cost. Value-of-time for a person is more complex. Not only is it hard to compute on a unit cost basis, but it is hard to compute on the basis of how much is enough time to make a difference. If 1,000 people coming in on Route 2 each save one minute—that is 1,000 minutes—do we simply take some aggregate value for the value of a minute of people traveling on that link and multiply it by 1,000 and use that as the economic value of having saved that minute? One could argue that, in pure economic terms, it makes sense, but one minute, or maybe even five minutes, has not really changed anybody's life in any substantial way. At what point is a bundle of time large enough to make a difference to an individual traveler?

How do we aggregate very small savings over very large numbers of people and make a judgment as to whether or not those are meaningful in decision-making?

Also valuing the "reliability" or variability in trip time for travelers is complex. Again, using inventory theory in freight, that variability was used to estimate the probability of a stock-out. The analog for travelers is not straightforward.

Other LOS Variables: Cost

There are other level-of-service variables. There is cost. Here, automobile costs can be complicated. There are "out-of-pocket" costs such as gasoline. Also, there are a number of fixed costs that do not vary much—or at all—with actual travel, for example, ownership costs. People often view these as "sunk costs." When they compare the costs of using the auto versus another mode, they treat the ownership costs as zero. In fact, it may be significant. Once you own the car, there is an economic incentive to use it. For other modes, cost is usually more straightforward. How much did the airplane ticket cost? How much did the bus ticket cost?

Service Frequency

Another level-of-service variable is frequency of service; this maps, directly or indirectly, depending upon the particular situation, into some kind of waiting time variable. One of the advantages that automobiles have is essentially an infinite frequency.

Waiting Time

Waiting time can be subtle. For example, consider a traveler who is catching the hourly flight from Boston to Washington. To say that the average waiting time for that traveler is a half hour, on the assumption that people arrive uniformly and independently of what the schedule is, is not reasonable. People schedule their arrivals based on a known departure time. An average waiting of 15–20 minutes is probably more reasonable. On the other hand, for a bus service that runs every five minutes, an average waiting time of 2.5 minutes is probably a reasonable assumption to make; people would not fine-tune their arrivals that carefully when the frequency is once every five minutes as opposed to once every hour.

Note also that, in the air case, service frequency matters above and beyond waiting time. Clearly, there is a convenience factor as well. If two airlines are providing service between two cities, the one with the greater number of daily flights is perceived as more convenient—"if I miss the 6:00 p.m. run, I will catch the 7:00 p.m.," rather than "if I miss the 6:00 p.m., I am stuck until 8:55 p.m." In air, the carrier with the higher service frequency has a larger market share; empirical data supports this.

Comfort

Comfort, broadly defined, is important. Is the bus air-conditioned? Is the food on the train edible? How do I value listening to Linda Ronstadt on the tape deck in my car?

Safety and Security

Safety and security are important LOS variables. Safety refers to the probabilities of accidents and their consequences. Security refers to car jacking or being mugged on a public transportation system. People respond to safety in different ways. While traveler fatalities in the air mode are well below that for automobile—the most dangerous part of air travel is driving to the airport—people do not perceive it that way.

In their car, they are in control. On an airplane, they are passive. Indeed, if one is sober, auto safety improves substantially!

Intangibles

In addition, we have many intangibles—some people like to drive cars. Conspicuous consumption and status in society are important. One cannot eliminate this from the equation, however hard it may be to quantify.

Mode Choice

The above is a reasonable set of level-of-service variables for travelers. Now, given these level-of-service variables, think about the choice of a transportation mode. Mode choice for travelers is a function of purpose of the trip and the modal options that are available to you.

Trip Purpose

Trip purpose is often categorized as follows:

- Work;
- Shopping;
- School;
- Pleasure;
- Personal.

Empirical data shows that people think differently about mode choice for different purposes. They make a mode choice as a function of trip purpose.

Modal Options

Next, consider modal options. Your modal options depend on your origin and your destination; you cannot take a choice that is simply unavailable. Land-use patterns constrain travel choice. We do not build subways in sparsely-settled Wyoming.

Options include private automobile, taxi, bus, train, boat, various intermodal combinations, bicycle, walking, and so forth. One can subdivide auto into single occupancy, carpools, and vanpools. So there are a variety of options, but your options are shaped by where you are and where you want to go.

In evaluating these modes, we have to think about the level-of-service that is being provided. But, recognize again: the level-of-service provided is often a function of the volume that is being carried by the particular mode. A decision that you make will have modest impact on level-of-service provided to others. But the decisions that we all collectively make have a great impact on the level-of-service that we observe on particular modes. This relates to the equilibrium concept. As level-of-service deteriorates because of high volumes, fewer travelers choose that mode. So, for example, as buses are subject to crush-loading, level-of-service deteriorates and people may make other choices, if they are able.

Different people make different mode choices. Income is a key variable in predicting how individuals may choose to travel. Car ownership is obviously a critical variable. Also, one has to be aware of the nature of the household or living group within which the individual making a travel choice lives. Household size, the ages of the people within it, the number of people that have jobs and their schedules—all of these are variables that will affect an individual's modal choice.

So, to summarize—why one is traveling, what modal options are available and the level-of-service of these options, and the characteristics of the individual making the choice, including living arrangements, are all central to mode choice.

We often see publicity campaigns to convince travelers not to drive to work and instead to use public transportation. It is argued that this has some benefits for society-at-large in terms of air-quality, and so forth. An interesting side effect occurs when people leave their cars at home, and other people in that home use that car. Just because you have managed to convince somebody to take transit rather than take his or her car to work, you have not guaranteed that car will simply sit in the driveway all day while that person is at work. Rather, somebody else in the household may be using it and, in fact, may use it much more intensively that it would have been used simply by driving in to work and back. So, the familial side has to be considered in transportation analyses.

Hierarchical Decision-Making

Transportation decision-making is hierarchical in nature. One does not wake up every morning and say, "I wonder how I'm going to get to work this morning?" One makes this decision in a more systematic way over a longer period of time. There is a good treatment of this in Ben-Akiva's and Lerman's text on demand modeling [2]. They talk about the hierarchical process of transportation choices; in their typology they describe long-range, medium-range, and short-range choices that people make about transportation and related activities.

Long-Range Choices

Long-range choices include employment—where am I going to work; residence—where am I going to live relative to where I work—as judgments that one makes "strategically." People who buy homes typically stay in them for some years. People who lease apartments typically stay in them for 12 months or more. People make job choices on a long-term basis.

Medium-Range Choices

In the medium range, Ben-Akiva and Lerman consider automobile ownership and mode choice to work. In the medium range somebody may decide to own a car and to drive to work most of the time.

Short-Range Choices

In the short range, people decide about route choice on a particular day. Also, in the short range, Ben-Akiva and Lerman include nonwork travel; that is, judgments that one makes about travel to shopping and other ad hoc activities other than the traditional journey to work. They include decisions about trip frequency, how often you travel, and the particular destination you may choose. Some mode choices may be short-range judgments as well, particularly with respect to nonwork travel, where the availability of a mode—that is, is the car home or not? —may affect the choice of mode.

This is a useful framework that they and others in the field have put together to develop models of transportation demand—how one goes about predicting what demand for transportation will be. The fundamental insight is that people do not make a single instantaneous judgment about trip-making and mode choice; rather one has to model transportation demand by thinking hierarchically about how long-, medium-, and short-range decisions lead to decisions about individual trips.

REFERENCES

1. Gakenheimer, R., "Urban Mobility in the Developing World," Transportation Research Part A 33, Pergamon, 1999, pp. 671–689.

2. Ben-Akiva, M. and S. Lerman, *Discrete Choice Analysis: Theory and Application to Travel Demand*, Cambridge, MA: The MIT Press, 1985.

CHAPTER 24

Intelligent Transportation Systems

....
Introduction[1]

In this chapter, we describe intelligent transportation systems (ITS). ITS combines high technology and improvements in information systems, communications, sensors, and advanced mathematical methods with the conventional world of surface transportation infrastructure. In addition to technological and systems issues, there are a variety of institutional issues that must be carefully addressed. Substantial leadership will be required to implement ITS as an integrator of transportation, communications, and intermodalism on a regional scale.

....
History

In 1986, an informal group of academics, federal and state transportation officials, and representatives of the private sector, began to meet to discuss the future of the surface transportation system in the United States. These meetings were motivated by several key factors.

First, the group was looking ahead to 1991 when a new federal transportation bill was scheduled to be enacted. It was envisioned that this 1991 transportation bill would be the first one in the post-Interstate era. The Interstate System, a $130 billion program, had been the centerpiece of the highway program in the United States since the mid-1950s.

---

1   Parts of the chapter are adapted from "ITS: A Short History and a Perspective on the Future," *Tranportation Quarterly 75th Anniversary Special Issue*, ENO Transportation Foundation, Inc., Lansdowne, VA, December 1996, pp. 115–125.

By 1991 this project would be largely complete. A new vision for the transportation system in the United States needed to be developed.

While the Interstate had had a major and largely positive impact in providing unprecedented mobility at a national level, transportation problems remained. From the perspective of 1986, highway traffic delays were substantial and growing. Rush-hour conditions in many metropolitan areas often extended throughout the day. Further, safety problems continued, particularly highway safety.

Also, the United States was concerned with the environmental impacts of transportation and the energy implications of various transportation policies. Any new initiatives in the surface transportation world had to explicitly consider environmental and energy issues.

Two more major motivations for considering the future of surface transportation were national productivity and international competitiveness, both closely linked to the efficiency of our transportation system. In 1986, America's major economic rivals in Western Europe and Japan were advancing very quickly in developing new technologies for use in advanced surface transportation systems. Their use of high technology concepts in the information systems and communications areas were seen as opportunities to revolutionize the world of surface transportation. This would improve the competitiveness of these nations and provide them with an important new set of industries and markets.

Further, it was recognized that these congestion, safety, environmental ,and productivity issues would have to be addressed largely by means other than simply constructing additional conventional highways. Particularly in urban areas, the economic, social and political costs of doing so were becoming too high. Thus, in 1986, this small, informal group saw before it an opportunity and a challenge based upon:

- New transportation legislation (at that time five years in the future);

- Concern for continuing transportation problems in the United States, despite major investment in the transportation system;

- The development by our economic competitors in Western Europe and Japan of various technologies that could enhance their industry posture and their productivity;

- Future limits on conventional highway construction, particularly in urban areas.

This informal group formed Mobility 2000 which eventually led to ITS America, a public-private organization designed to advise the U.S. Department of Transportation and advance the ITS agenda.

The essential concept was a simple one: marry the rapidly changing world of high technology with the world of conventional surface transportation infrastructure. The technological portion would include areas such as information systems, communications, sensors, and advanced mathematical methods. This concept could provide additional capacity with technological advances that could no longer be provided with concrete and steel. It could improve safety through technology enhancements and better understanding of human factors. Additionally, it would provide transportation choices for travelers and would control transportation system operations through advanced operations research methods.

What was envisioned and what came to be called intelligent vehicle highway systems (IVHS), and eventually intelligent transportation systems, is but another example of the marriage of transportation and technology as a phenomenon that has existed throughout human history. In the early part of this century, innovation in construction and manufacturing technologies made the current transportation system possible.

The ITS-4 Technologies

Now a new round of technological innovation, appropriate to the transportation issues of today, the "ITS-4," is here. These technologies deal with:

1. The ability to *sense* the presence and identity of vehicles or shipments in real-time on the infrastructure through roadside devices or global positioning systems (GPS);

2. The ability to *communicate* (i.e., transmit) large amounts of information more cheaply and reliably;

3. The ability to *process* large amounts of information through advanced information technology;

4. The ability to *use* this information properly and in real-time in order to achieve better transportation network operations. We use algorithms and mathematical methods to develop strategies for network control and optimization.

These technologies allow us to think about an infrastructure/ vehicle system, rather than independent components.

ISTEA

In December 1991, the Intermodal Surface Transportation Efficiency Act (ISTEA) became law. Its purpose was "... to develop a National Intermodal Transportation System that is economically sound, provides the foundation for the nation to compete in the global economy, and will move people and goods in an energy efficient manner."

As was envisioned in 1986, ITS was an integral part of ISTEA, with $660 million allocated for research, development, and operational tests. Additional federal, state, local, and private-sector funds were added to this initial allocation, leading to a substantial program.

The Strategic Plan

In June 1992, "A Strategic Plan for Intelligent Vehicle Highway Systems in the United States" was produced by IVHS (now ITS) America and delivered to DOT as a 20-year blueprint for ITS research, development, operational testing, and deployment [1].

The vision for ITS was articulated as:

- A national system that operates consistently and efficiently across the United States to promote the safe, orderly, and expeditious movement of people and freight. Here, recognition of the need to think intermodally and about the needs for both personal and freight mobility was explicit.

- An efficient public transportation system that interacts smoothly with improved highway operations. The concept that ITS had to do more than simply improve single occupancy vehicle level-of-service on highways is captured here.

- A vigorous United States ITS industry supplying both domestic and international needs. The plan noted that the United States market for ITS hardware, software, and services would be on the order of $230 billion over the next 20 years. Extrapolating this internationally, it is not unreasonable to think about a $1 trillion international market in ITS over that time period, well worth the effort for the private sector to pursue.

Technology, Systems, and Institutions

The plan focused on the triad of technology, systems and institutions. First, we have technology, the development and integration of technologies that would allow ITS to proceed. Second, we have systems, the integration of technologies into systems for operating ITS. Third, we have institutions, the challenges that face the ITS community in developing the public-private partnership and the government interactions that would have to be created at various levels. This third facet also addressed the educational challenge that the community faces and the organizational changes that would be necessary to have success in the ITS theater. It was recognized that while technology and systems were important issues in the development of ITS, the many institutional and organizational issues would be as complex and as difficult.

In the Strategic Plan, the recognition of the transportation/ information infrastructure was an important conceptual breakthrough. In other words, it accepted an intermingling of the new technologies in computers, communications, and sensors, with conventional infrastructure to create something wholly new in the world of transportation.

Functional Areas in ITS

It is convenient to think of ITS in terms of six areas.

Advanced Transportation Management Systems (ATMS)

ATMS will integrate management of various roadway functions. It will predict traffic congestion and provide alternative routing instructions to vehicles over regional areas to improve the efficiency of the highway network and maintain priorities for high-occupancy vehicles. Real-time data will be collected, utilized, and disseminated by ATMS and will further alert transit operators of alternative routes to improve transit operations. Dynamic traffic control systems will respond in real-time to changing conditions across different jurisdictions (for example, by routing drivers around accidents). Here are several important ATMS concepts.

Incident Management

The first is incident management. We are interested in reducing nonrecurring congestion; that is, incident- or accident-based congestion rather than rush–hour congestion. And it is clear that if one can identify and locate those incidents and remove them quickly, that one can reduce congestion–based delay substantially. With the technology on the network to measure velocities in real-time everywhere on the network, one can think about being able to identify and locate incidents by various velocity signatures.

Think about some major metropolitan area. There are only so many policemen and there are only so many television cameras to monitor the network. The system has hundreds of miles, and thousands of vehicles. It is easy to say, "locate and remove" incidents quickly. But where are they? And how do we deploy our emergency vehicles to get those incidents out of the way as quickly as possible? Having sensors in the ground that can measure velocities around the network enables the sensing of incident location in real-time. So ITS technologies, designed to monitor flow on the network and to allow us to apply various strategies using real-time information about flow, can also be used in the more specialized application of incident detection.

ITS is an enabling technology. We have technology put in place for a particular purpose—operating of the network—and as a result we have the ability to identify incidents or accidents.

Finding an incident is useful in and of itself, even if you do nothing about removing it quickly, because you can at least inform travelers of that loss of capacity and reroute them. You can change variable message signs, for example.

Electronic Toll and Traffic Management

Another valuable ITS concept is "electronic toll and traffic management" (ETTM). The basic idea here is that one can detect and identify individual vehicles with road-side readers; these readers can debit an account by sensing the transponder carried on your windshield; you pay your toll electronically rather than having to stop and pay. Here, the idea is that one can go through that toll at highway speed, 50, 60, 70 mph, and pay the toll. And that, in and of itself, reduces congestion.

Congestion Pricing—Revisited

Let us generalize the concept, though, and consider congestion pricing. If one can collect these tolls in real-time without stopping the cars, one has flexibility to control the traffic network both tactically and strategically. Once we have ETTM in place, we can change the toll as a function of time of day, for example. So one could have a low toll or a zero toll at off-peak hours and a substantial toll at rush hour.

This is what we discussed in the elevator problem, where we were concerned about peak loading on the elevator with everybody wanting to go to work at 9 o'clock; the way we handled that is we charged people for using the elevator at peak hours. By traveling off-peak, you pay less.

This economic construct has been talked about for half a century by transportation economists, and has been used in other businesses for many years. It is used extensively in the power industry and by the telephone companies, for example.

We now, through ITS and ETTM, have the ability to control network demand through pricing. Up to now our ability to do so in highway transportation was very limited. Now, because we can change the tolls in real-time through ETTM, we can implement congestion pricing. The tradition in the United States is that anyone can drive whenever they want, wherever they want, at no incremental cost beyond the cost they absorb due to congestion. If you want to go downtown at the worst possible time, it is a free country; it is your business. All you have to do is be prepared to take an hour rather than a half-hour to do it and burn the extra gas in the process. Of course, from an economic viewpoint, this driver does not absorb any of the externalities—the degradation in travel time that his entry onto the highway causes to everyone else, and the environmental impact caused by his vehicle, as well as the additional environmental impacts caused by other vehicles, for example. Congestion pricing can be a way to make that driver pay those externalities and hence be urged toward better behavior from a societal viewpoint.

The Philosophy of Highway Network Control

ITS allows us to rethink the philosophy of total freedom of mobility and raises the question of controlling the highway network. One can imagine controlling it with pricing; that is, setting prices in such a way as to evince particular behavior.

It will be very interesting to see how it develops from a political point of view, because it is really a fundamental shift in philosophy in how one operates the transportation system in general, and the highway system in particular. Congestion pricing, while economically sensible, may well not be politically viable in the United States.

Advanced Traveler Information Systems (ATIS)

ATIS will provide data to travelers in their vehicles, in their homes, at transit stations, or at their places of work. Information will include: location of incidents, weather problems, road conditions, optimal routings, lane restrictions, and in-vehicle signing. Information can be provided both to drivers and to transit users and even to people before a trip to help them decide what mode they should use. "May-day" calls for help from a disabled vehicle are included here as well.

Advanced Vehicle Control Systems (AVCS)

AVCS is viewed as an enhancement of the driver's control of the vehicle to make travel both safer and more efficient. AVCS includes a broad range of concepts that will become operational on different time scales. In the near term, intelligent cruise control, which automatically adjusts the speed of the vehicle to that of the vehicle immediately ahead, is an example of AVCS.

More generally, collision warning systems would alert the driver to a possible imminent collision, say, with a roadside obstacle. In more advanced systems, the vehicle would automatically brake or steer away from a collision. These systems are autonomous to the vehicle and can provide substantial benefits by improving safety and reducing accident-induced congestion. In the United States, this set of technologies is often referred to as the intelligent vehicle initiative (IVI).

In the longer term, AVCS concepts would rely more heavily on control that could produce improvements in roadway throughput of five to ten times. This concept is called the "Automated Highway System" (AHS). Movements of all vehicles in special lanes would be automatically controlled. One could envision cars running in closely-spaced (headways of less than one meter) platoons of ten or more, at normal highway speed, under automatic control.

ATMS and ATIS have already been applied in urban and suburban areas. AVCS, particularly AHS, is envisioned as a longer term program; indeed, AHS has had technical success, but suffers from political difficulties due to high costs, difficulty in demonstrating benefits, and the afore-mentioned long-term aspect.[2]

In addition, CVO, APTS, and ARTS are three major applications areas that draw upon ITS technologies.

Commercial Vehicle Operations (CVO)

In CVO, the private operators of trucks, vans, and taxis have already begun to adopt ITS technologies to improve the productivity of their fleets and the efficiency of their operations. Such concepts as weigh-in-motion (WIM), preclearance of trucks across state boundaries, automatic vehicle location for fleet management, and on-board safety monitoring devices, are included here. This is proving to be a leading-edge application because of direct, bottom-line advantages. Also, given the premium on productivity, technologies like ETTM are of special value to commercial fleets.

Advanced Public Transportation Systems (APTS)

APTS can use the above technologies to greatly enhance the accessibility of information to users of public transportation, as well as to improve fare collecting, scheduling of public transportation vehicles, intramodal and intermodal connections, and the utilization of bus fleets.

---

2   Indeed, in 1998, TEA-21, the Transportation Efficiency Act for the 21st century, eliminated most AHS funding.

Advanced Rural Transportation Systems (ARTS)

How ITS technologies can be applied on relatively low-density roads is a challenge that is being undertaken by many rural states. Safety rather than congestion is the main motivation for ARTS. Single vehicle run-off-the-road accidents are a target here. May-day devices are of particular interest in this environment.

Table 24.1 summarizes the subsystems of ITS.

TABLE 24.1   ITS Subsystems

		CHARACTERISTICS
ATMS	Advanced Transportation Management Systems	Network management, including incident management, traffic light control, electronic toll collection, congestion prediction and congestion-ameliorating strategies.
ATIS	Advanced Traveler Information Systems	Information provided to travelers pre-trip and during the trip in the vehicle. ATMS helps provide real-time network information.
AVCS	Advanced Vehicle Control Systems	A set of technologies designed to enhance driver control and vehicle safety. This ranges up to Automated Highway Systems (AHS), where the driver cedes all control to the system.
CVO	Commercial Vehicle Operations	Technologies to enhance commercial fleet productivity, including weigh-in-motion (WIM), pre-clearance procedures, electronic log books, interstate coordination.
APTS	Advanced Public Transportation Systems	Passenger information and technologies to enhance system operations, including fare collection, intramodal and intermodal transfers, scheduling, headway control.
ARTS	Advanced Rural Transportation Systems	Mostly safety and security technologies (e.g., May-day) for travel in sparsely-settled areas.

A Broad Systemic Approach

ITS represents a broad systemic approach to transportation. ATMS represents overall network management; ATIS is the provision of information to travelers. AVCS is a new level of control technology applied to vehicles and infrastructure. Applications in urban and rural areas involving public transportation, commercial vehicles, and personal highway vehicles, are encompassed by ITS.

There are important technological issues to be considered, many involving the integration of various hardware and software concepts on a "real-world" transportation network. However, few technological "breakthroughs" will be needed.

Institutional Issues

Of equal importance to technological and systems issues are various institutional issues that must be addressed if ITS is to be successfully deployed. Several are discussed below.

Public-Private Partnerships

A primary issue is the need for public-private partnerships for ITS deployment. One can contrast ITS with the Interstate System, the major transportation program in the United States in the twentieth century. The Interstate System could be characterized as a public works system. The funding was exclusively provided by the public sector and the fundamental decisions about the deployment of the Interstate System were made by the public sector.

ITS, on the other hand, will require deployment of infrastructure, largely by the public sector, and in-vehicle equipment by the private sector. Therefore, ITS can be characterized as both a public works and a consumer product system. This will require unprecedented levels of cooperation between the public and private sectors if ITS is to work effectively as a seamless national system. The hardware and software in the infrastructure must be compatible with the hardware and software in the privately-acquired vehicle.

While stand-alone ATMS (i.e., infrastructure) and ATIS (i.e., in-vehicle equipment) could work well, researchers are convinced that coordinated use of ATMS and ATIS will be much more effective than stand-alone systems of either type. Therefore, for optimal system operations, coordination and compatibility between ATMS and ATIS are essential. This requires close cooperation between the public and private sectors. In the United States, this cooperation has often not been strong. So ITS presents an important set of institutional challenges in developing an effective public-private partnership for ITS research and development, testing, and deployment.

Organizational Change

A second institutional question is the need for organizational change brought about by ITS. For example, our state Departments of Transportation have been based, for many decades, upon the technology of traditional civil engineering. Highway construction and maintenance have been the charter of state DOTs and, in fact, they have built a highway system that is unrivaled in the world.

However, that world is changing with social/political/economic constraints and with ITS coming on the scene. Now, rather than dealing with the conventional civil engineering technologies of structures, materials, geotechnical engineering, and project management, state DOTs need to be concerned with electronics, information systems, communications, and sensors. DOTs will need to emphasize the operation of the transportation system as well as construction and maintenance.

This is a fundamental shift for these public organizations. They will have to make a difficult transition over the next several decades for ITS to be successfully deployed around this nation, as will private sector organizations that have supported the historical mission. A whole new set of professionals will need to be attracted to these public-sector organizations and related private-sector organizations. In addition, fundamental changes in the missions of these organizations must come about.

It is interesting to observe that on the Central Artery/Tunnel program in Boston, Massachusetts, one of the last major projects of the Interstate System, ITS is playing a major role. Bechtel/Parsons, the contractor on the project, is putting considerable resources into understanding and developing ITS systems that can be used in conjunction with the development of conventional infrastructure to make sure this mammoth ($10 billion) mega-project will, in fact, work. Together with the Commonwealth of Massachusetts, the Massachusetts Institute of Technology, and MIT's Lincoln Lab, Bechtel/Parsons is working on traffic control centers, algorithms for effective routing of traffic, and roadside infrastructure that will permit efficient monitoring of traffic and incident detection.

The symbolism is strong. One of the great international construction consortiums working on the last of the great Interstate projects in this country is focusing on ITS technology to enable the finished project to operate effectively.

Transportation and Change

The linking of conventional infrastructure with the technologies of information systems, communications, sensors, and advanced mathematical methods for the movement of both people and freight is an extraordinary development. We cannot begin to foresee the changes (possibly both positive and negative) that will result from the development of this transportation/information infrastructure.

Changes Resulting from the Interstate

Think, for example, about the changes that came about as a result of the Interstate System, a $130 billion program, starting in 1956. The Interstate program can be thought of as an expansion, in-kind, of a conventional highway system. Granted, the Interstate was a substantial expansion in capacity and network size, but it was an in-kind improvement nonetheless. Yet, we had a hard time predicting what would happen as a result of this implementation. For example:

- The intercity trucking industry expanded and a financial blow was dealt to the railroad industry, as it lost substantial market share in high-value freight. This led, in turn, to a fundamental redefinition of the relationship between the public and private sectors in the freight industry in 1980, through substantial deregulation.

- The Interstate led to an unprecedented and unequaled mobility between and into United States cities and gave rise to the regional transportation concept, with wholly new methods of planning being required for region-wide analysis and design (e.g., Metropolitan Planning Organization [MPOs]).

- The Interstate System included the development of circumferential belts around major cities, leading to development patterns quite at variance with the ability of public transportation to service it and, as described by authors such as Joel Garreau, the development of "edge cities," a fundamentally new kind of urban structure [2].

- The Interstate led to a fueling of the post-war economic expansion and a period of unprecedented prosperity in the United States.

- A "stop the highway" backlash in urban areas resulted from the Interstate, and a political polarization between the build versus no-build factions became a fact of political life in United States transportation.

And all this resulted from an expansion, in-kind, of the highway system.

Changes Resulting from ITS

Regarding ITS, we have already seen:

- The reinvention of logistics through supply chain management, linking inventory management and transportation in wholly new ways;

- Dramatic moves into surface transportation by organizations not traditionally involved, such as the national labs and aerospace companies in the United States;

- Changes to academia, with new alliances and new academic programs beginning to be formed, and faculty participating in transportation education and research who have never been part of that process before;

• Building of new relationships among public-sector agencies to enable regional and corridor-level system deployment.

These have already happened and it is just the beginning. We cannot begin to foresee all that will occur. The enabling technology of ITS—the transportation/information infrastructure—can and will have profound effects. We hope they will be positive—accessibility, economic growth, improved quality of life, improved information for planning, and intermodal transport. However, unforeseen outcomes—both positive and negative—are certain with this new transportation enterprise.

The Post-Strategic Plan Period

The years since the strategic plan have been busy ones in the ITS community. Program plans that translate the Strategic Plan into specific shorter-term actions have been developed. A National ITS System Architecture has been developed and regional architectures are being developed. The United States DOT has established the Joint Program Office as a group that cuts across the modal administrations of DOT to address ITS research, development, testing, and deployment. ITS America continues to grow, with more than 1,000 members. The international community in ITS is cooperating at a professional level with the ITS World Congresses, initiated in Paris in 1994, and continuing annually in Asia, Europe, and North America, most recently in Toronto in 1999. Turin will be the site of the first World Congress of the new millennium in 2000.

Successes are appearing. To name but several: one can look at TRANSCOM in the congested New York, New Jersey, Connecticut region for an example of an ITS deployment providing ATMS, ATIS, and electronic toll collection in the tri-state area. The SmarTraveler program in Boston, and other cities, is an example of an advanced traveler information system with a strong initial track record. The Houston public transportation system is yet another example of an ITS deployment which is quantitatively and qualitatively changing the supply of transportation service in the Houston metropolitan area. Deployments in Western Europe and Japan are advancing as well.

Private-sector organizations have continued active programs in ITS. The automobile manufacturers in the United States and abroad are marketing advanced traveler information systems, supported by major initiatives in communications and in computerized mapping.

Cooperation between various public-sector agencies also characterizes the ITS movement. Commercial vehicle operations initiatives, such as HELP and Advantage I-75, involve many public jurisdictions as well as private-sector truckers. The I-95 Coalition cooperates on ITS technologies stretching from New England to Virginia.

Efforts to make ITS truly intermodal, both for travelers and freight, are under way. For example, exploiting ITS to enhance truck-rail-ocean freight intermodalism is high on the agenda.

Regional Deployment: A Strategic Vision

The focus for ITS in the future is clearly on deployment. Taking research and operational test results and putting them into routine practice is the emphasis in the ITS world today. How to best advance the deployment agenda is currently a matter of intense discussion in the ITS community [3].

Some argue that the best approach is for ITS to focus on regions as critical units of economic competition, building on the idea that ITS enables deployment of regionally scaled transportation control strategies. Often, we speak of the "competitive region." The work of Rosabeth Moss Kanter [4] at the Harvard Business School emphasizes the idea that subnational units will compete economically on the basis of productivity and quality of life provided for its citizens. (A good deal of attention is being devoted to "regional architectures," developed "consistently" with the National ITS System Architecture, as a mechanism for working toward national interoperability [5].)

This regional concept can be combined with two others. First, the natural partnership between ITS and the nascent National Information Infrastructure (NII), a communications network of unprecedented scale, scope, and functionality, can provide substantial deployment benefits to both. Second, the strong trend toward freight and traveler intermodalism provides a critical boost to ITS technologies. This is where ITS can help overcome intermodalism's weak point at interfaces between modes through information and communications technology.

Pulling these ideas together,

The strategic vision for ITS, then, is as the integrator of transportation, communications, and intermodalism on a regional scale.

This is, of course, merely a prediction. What will transpire, only time will tell.

The following chapter returns to the topic of networks and ultimately relates network concepts to ITS.

REFERENCES

1. *A Strategic Plan for IVHS in the United States*, ITS America, Washington, D.C., May 1992.

2. Garreau, J., *Edge City: Life on the New Frontier*, Doubleday, 1991.

3. Sussman, J. M., "ITS Deployment and the 'Competitive Region,'" *ITS Quarterly*, Spring 1996.

4. Kanter, R. M., *World Class: Thriving Locally in the Global Economy*, New York: Simon & Schuster, 1995.

5. Sussman, J. M., "Regional ITS Architecture Consistency: What Should It Mean?," *ITS Quarterly*, Fall 1999.

CHAPTER 25

The Urban Transportation Planning Process and Real-Time Network Control

....
Networks

Individual people make individual decisions about how much to travel, where to travel, by what mode to travel, and by what path or route to travel on the system (there is a comparable set of questions about freight). From the point of view of understanding the need for transportation behaviorally, these individual decisions are important. But we are also interested in network behavior—how the network as a whole behaves—which involves going from choices that individual people make about transportation to understanding how the network in its entirety operates. What are the flows on the network? What are the levels-of-service being provided? How much capacity is needed at various links and nodes? What might be done to enhance the network? How can we control the network in real-time?

We need to aggregate predicted individual choices to overall network flows and performance. We begin with how this is classically done in an urban transportation context through what is called the "Urban Transportation Planning Process."

....
The Urban Transportation Planning Process

We begin the process by building a network which itself requires an aggregation process. In performing this kind of analysis, we take a set of geographical areas, and aggregate them into what are called "zones." For each zone, we define a node or centroid, often at the population center or major activity center of the zone. A network is constructed by connecting the nodes with links (see Figure 25.1). On this network, we overlay the transportation network, which is usually multimodal, including, say, highway (both auto and bus) and rail transit. Some judgment is required to do this overlay. For example, we may need dummy nodes (in addition to zone centroids) to reflect the topology of the transportation network (see Figure 25.2). Usually, we try to make those zones as homogenous as possible in terms of land use and economic characteristics. Zones both generate and attract trips.

Choosing the Number and Size of Zones

There is a fundamental modeling trade-off here. We can choose a relatively small number of large-area heterogeneous zones, which makes network analysis easy because, with a small number of zones, the network is simple. Alternatively, we can choose a large number of smaller homogenous zones with the characteristics of individuals within those zones more alike; however, we then have a large number of zones, and a network that may be too complex to effectively analyze. In either case, some kind of aggregation process has to take place.

FIGURE 25.1
Constructing a network from zones.

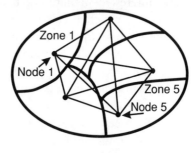

FIGURE 25.2 *Nodes as zone centroids and "dummies."*

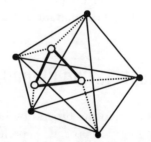

Within this process, well known in practice for at least three decades, there are four steps that one uses for modeling network behavior.

Trip Generation

First is trip generation—deciding whether to travel.

Trip Distribution

Second is trip distribution—once a decision has been made whether to travel, where is this trip destined for?

Mode Split

Third is mode split or mode choice, as it is sometimes called—how do people split up among the transportation modes available to them—highway, rail rapid transit, bus, and so forth?

Assignment

Fourth is assignment—for each of those modes, what are the routes that individual travelers take?

Let us do this step-by-step. We first assume that each zone has some kind of trip generation characteristics. It generates trips as a function of land use and the demographics of people within it. Each zone also has some attractive characteristic; that is, it attracts trips to it. For example, a zone with many employment opportunities (i.e., destinations) would tend to be a very attractive zone. The steps of generation and distribution estimate how many trips are made from zone i to zone j on the network. Once that is done, we next estimate how the travelers split among the transportation modes that are available; and finally, how those travelers are assigned to particular routes between zone i and zone j.

Let us look at this process. For simplicity, let us assume we have only one mode, so we skip the mode choice step. Further, we assume we have generated and distributed the trips between all pairs of nodes (i.e., our concentrated representation of our zones). We represent this with an origin-destination matrix, the $(i,j)^{th}$ element of which is the trip from node i to node j (see Figure 25.3).

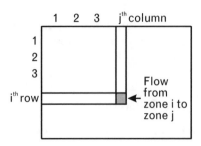

FIGURE 25.3
*Origin-destination
matrix.*

Now, assignment, estimating the flows on individual links is the next step; we are interested in how the origin-destination trips are distributed over various paths on the network. This can be posed as a mathematical program in which the times on the links are a function of the volumes on the links—level-of-service deteriorates as volume approaches capacity. Let us consider the user-equilibrium formulation of this problem.

User Equilibrium

User equilibrium is based upon the idea that each user in the system is trying to optimize his or her own trip through the system. There is no global optimization taking place. No overall network manager is making judgments for all travelers about how to distribute themselves on the network in order that the network behave optimally. Each traveler, in true micro-economic fashion, is trying to optimize his trip without concern for other travelers. User equilibrium is said to be achieved when all the paths that are used between any origin-destination pair all have the same travel time. Unused paths will have a larger travel time.

Consider, for example, trips between node 1 and node 2 (see Figure 25.4). One can go from node 1 to node 2 directly from node 1; one can also go from node 1 to node 2 via node 3. Another possibility is

FIGURE 25.4   *Flows
between node pairs.*

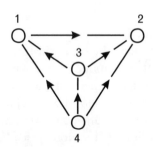

node 1 to node 4 to node 3 to node 2. And there are others. User equilibrium occurs when all the used paths between 1 and 2 (if that was the only flow) have the same travel time and all unused paths have a longer travel time. Simply put, no one gains by switching.

Now, usually, of course, there would be flows between many or all of the node pairs, not just one, and the above condition would have to be true for all node pairs for user-equilibrium to exist.

As noted above, the user-equilibrium concept can be formulated as a mathematical programming problem, which gives you flows on the links. We will not dwell here on this formulation. A numerical solution, called incremental assignment, is briefly discussed in Chapter 5.

System Equilibrium

Let us distinguish between the system-equilibrium and the user-equilibrium approaches. Imagine instead of everyone acting selfishly, we try to organize the flows on the network to minimize overall costs.

Suppose the total cost function for operating this network was the sum of the cost of traversing each link multiplied by the flow on each link. These link flows may be composed of several origin-destination flows. C_{kl} is the link cost on link k-l and F_{kl} is the link flow on link k-l (see Figure 25.5).

Now, we organize the flows so that the overall cost on the network is optimal, rather than each individual user optimizing their own route on the network. This is the system-equilibrium approach. If one were operating a transportation network, in which all the vehicles were under your control, you would not say to each driver, "Select your best route," but rather you would come up with a system optimum set of routes that minimizes the following equation:

$$\text{Min } C_T = \sum_{all\ k,l} C_{kl} F_{kl}$$

We emphasize that the link costs C_{kl} will be some function of the flow, F_{kl} (also called the volume), on the link. Often, the cost of traversing the link is simply the travel time of that link multiplied by some constant representing value-of-time. This travel time will go up as the flow F_{kl} increases. We can show that the optimal value of C_T in the system-equilibrium formulation will always be less than or equal to the total cost of operating the network in a user-equilibrium formulation.

FIGURE 25.5
Total costs.

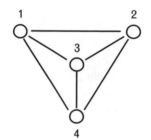

Total cost $= \sum\limits_{all\ k,l} C_{kl} F_{kl}$

The flow on link 3–2 may be composed of origin-destination flows 1–2, 3–2, 4–2, etc.

Network Planning

Now, what can one use these analysis tools for in network planning? One can try to understand what will happen if one modifies the network, as one works toward trying to satisfy transportation demand or one works toward a more effective way of providing transportation service. For example, using this kind of formulation, one could choose which links to upgrade. Upgrading a link makes it cheaper to use because travel times are reduced for a given volume (or flow).

So, in the planning context, you could look at an existing network and make judgments about strategic investments you could make to enhance transportation in the region. How do you best spend the dollars? Do you add a lane here, or do you maintain this link more effectively, or do we work on this bridge here, which has been a bottleneck?

So, we can experiment with our network models; add capacity to a link (or set of links) and see what happens to the flows. Was that investment cost-effective? We can try other possibilities and search for the best way to allocate scarce resources.

Now, realize so far we have been talking only about assignment—how are O-D flows assigned to paths? Now, consider the full four-step process. This includes estimating:

1. How many trips a zone will generate;

2. Where people want to travel (which zones are attractors);

3. How travelers choose among mode choices;

4. How traffic is assigned to paths and links on the network.

Thus far, we have discussed only the last step. If we are doing a complete network analysis, we would need to consider the other three steps as well, but we will not discuss those steps here.

Networks and ITS

Now, thus far we have considered a planning perspective: thinking strategically about deploying resources and investing in a transportation network. But let us consider ITS again. Network control is a key element of ITS. Advanced Transportation Management Systems (ATMS) manage traffic networks in real-time.

Real-Time Network Management

Now, we are no longer talking about expanding a bridge from two lanes to four lanes, which takes place over several years—a strategic investment in the transportation network.

Now we are considering ITS as a control mechanism for tactically controlling the transportation network in real-time. We are interested in what happens between 8:00 and 8:15 this morning and operating the network in an effective way in that time frame.

Think about how you might go about doing that in this same network framework. In planning, one typically will have an origin-destination matrix, say for daily traffic—the number of trips we expect to go between this origin and this destination by this mode on a particular day. Possibly one has an origin-destination matrix for the peak hour to understand how the system would operate during rush hour when it is most stressed.

In order to have real-time ITS applications, one has to think more subtly about how these networks behave. For example, one would need to do a traffic assignment in this real-time environment; rather than having an origin-destination matrix for the entire day or for the rush hour, one might have an origin-destination matrix for the network for every 15 minutes. You would need a prediction of origin-destination flows between all point pairs on the network between 7:00 and 7:15, 7:15 and 7:30, 7:30 and 7:45, and so on. So in principle you could predict—given your time-varying origin-destination matrices and network assignment technique—in real-time, where congestion on the network would occur. And you might—given real-time information such as the sort that ITS provides—be able to make control decisions that would allow you to optimize the flows on the network in real-time. You have strategies, such as changing speed limits, changing ramp-metering rates, providing routing instructions via variable-message signs or via in-vehicle displays, changing traffic signals, and so forth, available to you to control flow to some extent.

Now, we have discussed two ways of using the network framework. One is planning-oriented and strategic; one is operations-oriented and tactical; both are based on the same fundamental network approach.

Why the Tactical Problem Is Hard

The tactical problem is very difficult. This dynamic network equilibrium problem is an important research area. There are a number of people around the world working on this dynamic traffic assignment and real-time congestion prediction problem.

Why is there so much interest? This has always been an interesting mathematical problem but, now, it has become an interesting practical problem as well, because we have the ITS technologies that allow us to monitor flows in real-time, which allows us to deploy control strategies so that we can optimize how a system operates in real-time. This is a very tough computational problem and we merely scratch the surface here. However, we can see the conceptual view. We are entering simulated traffic into the network at 15-minute intervals and merging it with the real traffic; we are measuring flows on individual links, recognizing that the time to travel on those links is a function of the volumes on those links; and we are trying to come up with control strategies that will allow us to optimize the network.

This is a complex problem to solve, but it is made even more complex by the fact that, if the solution is going to be useful, it has to be produced in real-time. The methods need to tell us how to control the network in real-time.

Let us say we have a huge network that we are studying for (strategic) planning purposes. We are deciding whether to make that bridge four lanes, or we are deciding whether to expedite maintenance on some other link; we have many alternatives in a complicated network. It is not a critical problem computationally because you are operating on the scale of weeks and months and years in terms of making these decisions. So you let the computer run all weekend, if you need to, to figure out three or four alternatives. You look at the information that comes out and you hope it will give you some guidance on planning the network. And the fact that it used a lot of computer time is a second-order consideration.

On the other hand, suppose you are looking at the real-time operating problem, pumping traffic into the network every 15 minutes with different origin-destination matrices, with the idea being to test out various strategies for operating the network; if it takes you a half-hour to figure out what you ought to be doing in the next fifteen minutes in real-time, that is not very helpful. It may be an elegant mathematical result, but from the point-of-view of deciding what to do—should we divert traffic from Route A to Route B? —if it takes us a half-hour to figure out what to do in the next fifteen minutes, it does not help. So, not only is it computationally very complex; it is also a real-time problem which makes the constraints on what we can do all the more difficult.

Ashok Formulation

Now, we show in Figure 25.6 a flowchart that illustrates these ideas.

Start in the upper left-hand corner with traffic on the network. The surveillance system monitors the network with various sensors; that surveillance information is abstracted and input into what Ashok calls "network state estimation"; in his research, he considers various ways of representing the state of a network. From that estimation of the network state—now in the lower right-hand corner—he develops various control strategies for the network: what do you do to make things better in real-time on this complex network? So, he considers various strategies; then he executes his network state prediction module.

You know what the state of the network is; you also have an estimate of origin-destination demand in the immediate future. Ashok lists several operating strategies: compute percentage of traffic to divert (to another route), change the use of particular lanes, ramp-metering, and so forth.

So he has these various control strategies he can use to make the network operate better. So he asks, "If I execute a particular control strategy, what is going to happen on the network?" The "network state prediction" box answers this question with rather complex, mathematical procedures.

Once he makes those predictions he asks, "How well did the strategy perform?" How well does that particular strategy work with respect to various measures of effectiveness (MOE): travel time, emissions, and so forth? Then finally the module that predicts the

FIGURE 25.6   *Traffic control system functional structure (Source: Ashok, K., "Estimation and Prediction of Time-Dependent Origin-Destination Flows," Ph.D. Thesis, Department of Civil and Environmental Engineering, Massachusetts Institute of Technology, September 1996.)*

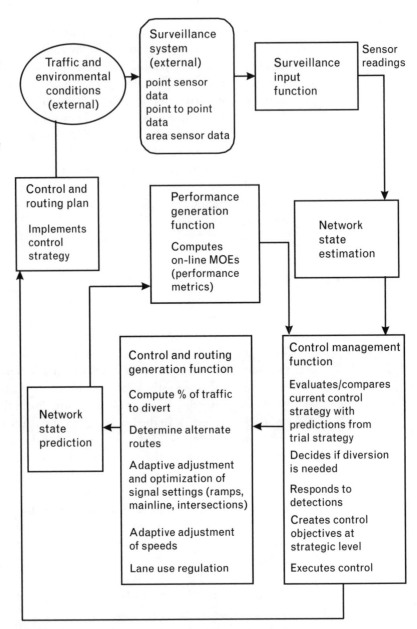

performance of various operating strategies goes back into the control management function and ultimately implements a control plan. We choose the best strategy and implement it. Then we start the whole process again, maybe five minutes later.

Now, this is a pretty complex, logical structure. You can get an intuitive feeling for that. But bear in mind again that it has to be done in real-time. It does not help if it takes more time to do the network state prediction and calculate the MOEs for the various alternatives than the time frame for implementing strategies on the physical network we are trying to control.

Some Research Ideas

Now, here is an interesting question. How much is it worth to have better information about how the network is operating?

Value of Perfect Predictions

In statistical decision theory there is the concept of the value of perfect information. How useful is information for making decisions? The question here is how much research energy should we put into sophisticated methods for predicting how networks will perform. It depends upon how good we think we are in coming up with good control methods. That is, if we don't have good ideas for strategies to use in optimizing network flows, it probably doesn't make sense to expend a lot of computational and research energy to compute the future state of the network. If we cannot use those predictions to implement reasonable strategies for making traffic better, then how useful are the predictions anyway?

The Link Between Prediction and Control Methods

If one does not have a good use for the predictions in real-time because you do not have good strategies yet, maybe the information is not worth developing. This is the kind of question that we ask both in practice and in research: Is the information that we are developing going to be useful to us in actually operating our system better?

Formal Problem Statement

To be more formal about operational strategies and predictions, suppose we can estimate the current state of the network perfectly, and also predict perfectly what will happen in 10 minutes, given an operating strategy.

Using those perfect estimates and predictions, we can select the network operating strategy from those available. Let us call that NP_{pp}, which we define as network performance with perfect predictions. In this case, this is constrained only by how good we are at thinking up operational strategies.

Suppose NP_{P_j} and NP_{R_j} are predicted and real (actual) network performance using strategy i (see Figure 25.7).

In this case with perfect predictions NP_{P_1} equals NP_{R_1} and NP_{P_2} equals NP_{R_2}. The reality of what happens in 10 minutes is exactly what we predict and, in this case, since option 1 is better than option 2, NP_{pp} equals NP_{P_1}. Now, of course we do not have a perfect prediction, as shown in Figure 25.8.

So, using that (imperfect) prediction, we select an operating strategy and again measure network performance.

Now, there are several different directions research in this area could take.

Research Direction 1

Assume perfect prediction—then work on developing better operating strategies. That means working on operational strategies that improve NP_{pp}.

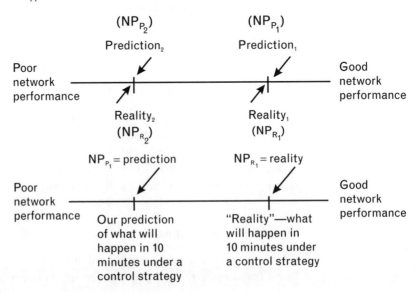

FIGURE 25.7
Prediction/reality.

FIGURE 25.8
Imperfect prediction.

Research Direction 2

This research would deal with understanding the sensitivity of network performance to our ability to predict. Here we would compute the difference between NP_{pp} and NP_{ip} as a function of ["reality"−"prediction"], where NP_{ip} is network performance with imperfect prediction. In theory, the larger the difference between reality and prediction, the poorer NP_{ip} will be. This is illustrated in Figure 25.9. In the above case, we would select strategy 1 when we should have selected strategy 2 because of our difficulties in making predictions.

$$NP_{ip} = NP_{R_1}$$

$$NP_{pp} = NP_{R_2}$$

The resulting loss in performance because strategy 1 was incorrectly chosen is $NP_{R_2} - NP_{R_1}$.

Research Direction 3

Work on improving predictions. In our terminology, this means to minimize the difference between reality in 10 minutes and our predictions of what will happen in 10 minutes. In equation form, this is as follows:

$$Min \; \{ \; Reality - Prediction \}$$

$$Min \; \{ \; NP_{R_1} - NP_{P_1} \}$$

FIGURE 25.9
*Selecting the wrong
strategy.*

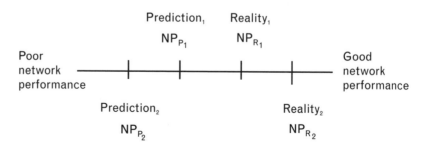

Now, if it turns out that under research area (2), that NP_{R_2} is almost equal to NP_{R_1}, then perhaps it does not pay to do much on research area (3) until research area (1) produces a new set of operating options that can take better advantage of better predictions. This is shown in Figure 25.10.

The fact that our predictions are not very good in the circumstances shown above is not important because the real conditions do not vary very much as a function of the strategies available to us for improving network performance. Our prediction methodologies—for the moment—are not very important. What we really need to do is focus on coming up with better operating strategies, such as strategy 3 below, which is "much better" in reality than strategies 1 and 2. That will make better predictions worthwhile. Note that in this case, we would choose strategy 1 (because our prediction says it is the best) although it is the worst strategy, and indeed "much worse" than strategy 3, although only modestly worse than strategy 2.

Now, recall that all of this is going on in real-time. None of it makes any sense at all if we are unable to run these algorithms fast enough to make decisions in a timely fashion in the context of our real transportation network.

FIGURE 25.10
Importance of prediction quality.

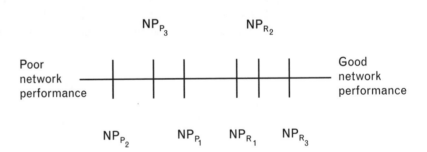

CHAPTER 26

Traffic Signals and Other Control Measures

Traffic

Next we will discuss various aspects of highway traffic. We first discuss traffic lights and traffic light synchronization and other traffic control measures.

Traffic Light Synchronization

When we consider traffic light synchronization, a useful tool is a space-time diagram (see Figure 26.1). This is a plot of the location over time for a vehicle. Suppose three traffic lights are spaced along an arterial at points A, B, and C. The green band is a time-space trajectory through which a vehicle can go and not be stopped by a red light.

The slope of the line defining the green band is speed (the ratio of distance to time). If a car stays within that green band as it goes through traffic signals A, B, and C, it will continue unopposed by a red light. The theory is that we can synchronize consecutive lights to minimize delays so that people can proceed from origin to destination with a minimum number of stops.

Now, of course, this is not straightforward. We have traffic going in the orthogonal direction as well. Suppose we have a grid network; the question is how one builds the system considering flows in both directions (see Figure 26.2).

This is a hard problem. We consider the design of splits—dividing the total cycle time (the time between the start of consecutive reds) between the red and green. We consider offsets—the time between

FIGURE 26.1
Space-time diagram.

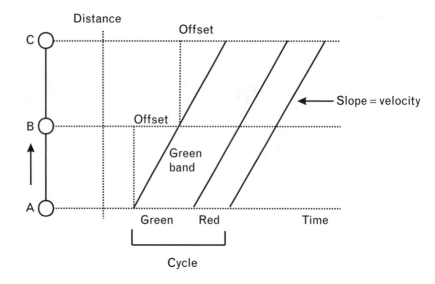

FIGURE 26.2
Grid network.

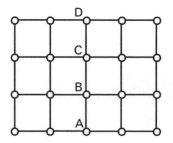

light A and light B turning green in Figure 26.1. The question is how to design the splits and the offsets in the two directions optimally.

Optimizing Traffic Light Settings

Even the question of what optimally means is not straightforward. There are a number of measures one can optimize. For example, one might consider the total amount of time spent stopped at red lights for vehicles in the system, considering both directions. An alternative measure is the number of times that individual vehicles need to stop. The notion here is that we are willing to live with a certain amount of delay in the system, but what people really want to do is keep moving along. And, of course, simply optimizing the total time in the system is another approach. Often, people use weighted methods of these different parameters based on their experience and professional judgment.

Traffic light synchronization is made more complex by other vehicles and congestion, people wanting to make left turns, and pedestrians—all the things that happen out there in the real world. So, synchronization—in reality—is difficult.

Traffic Light Synchronization: Levels Of Sophistication

There are levels of sophistication for synchronizing traffic lights. They range from virtually no attempt to synchronize lights to rather advanced methods. At the low end of the scale, there are traffic light systems that make no attempt to coordinate traffic signals; each traffic signal on the grid operates independently.

The Minus-One Alternative: Mystic Valley Parkway

While nonsynchronization may be the zero alternative, the minus-one alternative exists as well. In neighborhoods where the residents are trying to deter traffic, you may see a sign that says "Signals Timed to Require Frequent Stops" (see Figure 26.3). The lights are, in some sense, coordinated to maximize, not minimize, the travel time through the system.

Now, let's discuss some progressively more sophisticated ways of doing light synchronization.

Static Synchronization

At a first level, we can synchronize statically; there is one setting for the system all the time. We optimize for typical traffic and the system operates on that setting all the time. So, perhaps you encounter a long red at midnight, at a signal optimized for rush-hour flows.

Time-of-Day Settings

At the second level we have time-of-day settings. We recognize that the rush hour is different than noon or 7:00 p.m. or 2:00 a.m. So, we use different cycles at particular times of day by developing a set of static plans to be used at different times of the day.

FIGURE 26.3
Street sign.

Predefined Plans

At the third level are predefined plans that can be deployed as a function of traffic conditions. One can design plans that will operate effectively under particular traffic conditions. We sense traffic conditions—for example, heavy congestion on Memorial Drive—with detectors of some kind. Now, having sensed traffic conditions and knowing time-of-day—which is, in effect, a prediction of the near-term future—one could use a pre-stored algorithm for that particular condition. So this introduces some degree of dynamism—the idea that one can think through seven or eight plans that would be effective under seven or eight different kinds of traffic conditions. When you sense those traffic conditions, you simply use the appropriate plan.

Dynamic Systems

The fourth and ultimate level is the idea of totally dynamic systems—dynamic real-time traffic light settings that are computed in real-time as a function of traffic conditions, as well as some prediction of upcoming volumes. For example, a car breaks down on Memorial Drive. The traffic is backing up and flowing over onto Magazine Street as people try and take a back way through Cambridge. Let's give a little more green to Magazine Street so people can get to Harvard Square more quickly.

So, the ultimate level, with current technology, is a dynamic real-time setting for all signals computed as a function of what is happening on the network; we measure volumes, speeds, and queues and predict near-term future traffic entering the network. You integrate those predictions with what is actually occurring on the network; you do some optimization that synchronizes your traffic settings.

Now, of course, to do all this, we need real-time traffic information flowing into a computer system and algorithms computing what to do next, again in real-time.

Gating and Draining

Two of the techniques used in dynamic systems are gating and draining. If congestion exists on a particular link (because a major employer is letting out employees, for example), we can gate the approaches to that link with extended reds and drain the area with extended greens (see Figure 26.4).

Other Traffic Control Ideas

Here are some other measures that are useful for traffic control in particular in urban areas.

FIGURE 26.4   *Gating and Draining*.

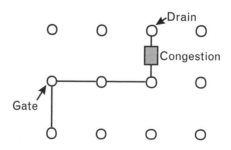

Ramp-Metering

First, there is ramp-metering. Ramp-metering consists of a dynamic traffic signal that controls the access from surface streets onto limited-access expressways. The ramp-meter, by changing from red to green, can regularize the flows of traffic from the ramp onto the expressway and can smooth the flow on the expressway itself. Operating experience has shown that one can create a smoother, higher volume and higher velocity traffic flow on the expressway by regularizing the flow of cars from the ramps. Sensors measure the length of the queue at the ramp- meter and the characteristics of the traffic flow on the expressway and cars are allowed into the traffic stream as appropriate. Empirical data demonstrates benefit for the overall flows in the area, from a volume as well as safety viewpoint.

Reflecting on the way in which these concepts fit into the overall societal perspective on transportation systems, environmental impact statements have been required for deployment of ramp-meters. There are two environmental concerns raised about ramp metering. One is fairly obvious, which is that the people living near the ramp-meter are concerned with car queuing in their neighborhoods.

The other concern has been the traffic has been smoothed and therefore a higher volume flow has been achieved. The improved traffic flow will attract more volume—remember supply and demand. That means more tailpipes and that is environmentally unsound. This is a serious question; it goes to the issue of measures of effectiveness for transportation system performance. Is better flow good or bad? This is a political question.

Dedicated Bus Lanes

Second, are dedicated bus lanes—where traffic is limited only to buses—and again, here the theory is that one can provide a higher level-of-service for buses, and therefore attract people from their automobiles to take buses. Indeed, by taking a lane away from cars, level-of-service deteriorates for cars as well.

A problem in conventional bus systems is that they are in the same traffic stream as cars. When level-of-service goes down for automobiles, it is poor for the buses as well. If there is a dedicated lane, one can have a higher level-of-service for the bus than one has for the traffic at large. We see these dedicated lanes on expressways and also sometimes on surface streets.

There are often political issues associated with this kind of development. People feel as though their lane for autos has been taken away from them; balancing the common good with their own good is a troublesome issue for many people.

Reversible Lanes

Third, the idea of reversible lanes is a notion traffic engineers have been using for some years. Basically the notion here is as follows: on an arterial street which has six lanes, one could set the traffic signals so that there are four lanes for in-bound in the morning and two lanes out-bound. Then, for the evening rush hour you simply reverse the traffic signals, so your capacity is four lanes out-bound and two lanes in-bound. One can flexibly change the directional capacity of the system.

Earlier, we noted as one of the fundamental characteristics of transportation systems, the directionality of flows and flow imbalances. Everybody wants to come into Boston in the morning. Everybody wants to leave in the evening. Here is a way of adjusting capacity to recognize that kind of imbalance.

High-Occupancy Vehicle Lanes

Fourth, are high-occupancy vehicle lanes (HOV lanes). Here, we are trying to entice people to carpool—to get away from the single-occupancy vehicle (SOV) mentality. You have lanes dedicated to cars with three or more people. So people who can carpool with neighbors, or family who want to leave at about the same time can use a dedicated lane that presumably can get them to work more quickly. You can combine the last three ideas—make the extra rush-hour lane an HOV and bus lane as well. The issue with high-occupancy vehicle lanes is, again, political. People question why you are taking this lane away from the general public.

High-Occupancy Toll Lanes

A related idea is high-occupancy toll lanes (HOT lanes). Here, you can use the HOV with less than the minimum number of passengers, but you pay a toll. ITS techniques enable us to do this effectively.

Traffic Calming

Finally, there are so-called traffic calming techniques, intended to slow down traffic, usually in residential or school neighborhoods. Speed bumps are an example of a traffic calming method, as are frequent stop signs and traffic signals like the one in Figure 26.3.

CHAPTER 27

Deterministic Queuing

Deterministic Queuing Applied to Traffic Lights

Here we introduce the concept of deterministic queuing at an introductory level and then apply this concept to setting of traffic lights.[1]

Deterministic Queuing

We begin our treatment of deterministic queuing with a simple highway example. Assume a rate of flow of vehicles that is a function of time: $\lambda(t)$. This is deterministic flow; there is no stochasticity. The potential flow out of the system is a function of time $\mu(t)$; vehicles arrive and exit deterministically as well. In the first situation, we consider $\lambda(t)$ and $\mu(t)$, as shown in Figure 27.1. In this example, $\lambda(t)$ is unsteady flow; the rate of vehicles entering the system varies with time. Here, we see the peaking behavior that is characteristic of transportation systems. We also assume that the output capacity of this highway is constant over time; that is, it is capable of outputting a constant 1,500 vehicles an hour.

Now, obviously, the system does not always output 1,500 vehicles an hour; for example, for $0 < t < 1$, there is zero output because there is zero input. From $t = 1$ to $t = 2$, although the highway is capable of outputting 1,500, it outputs only 1,000 because that is all the vehicles there are. However, between $t = 2$ and $t = 3$, we start getting into some trouble, because while 2,000 vehicles arrive, only 1,500 can be served. We

1 A more detailed treatment of this material can be found in Mannering, F. and W. Kilaresky, *Principles of Highway Engineering and Traffic Analysis*, 2nd Ed., New York: John Wiley and Sons, 1998, an excellent overall reference text for highway and traffic analyses.

FIGURE 27.1
Deterministic arrival and
departure rates.

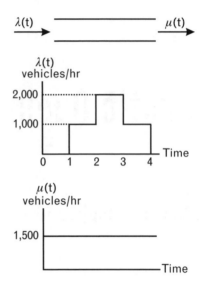

have to reconsider how we model that and how we measure the performance of a system of this sort.

....

Queuing Diagram

We study these systems through a construct called a *queuing diagram* (see Figure 27.2); on the vertical axis, the diagram represents the

FIGURE 27.2
Queuing diagram.

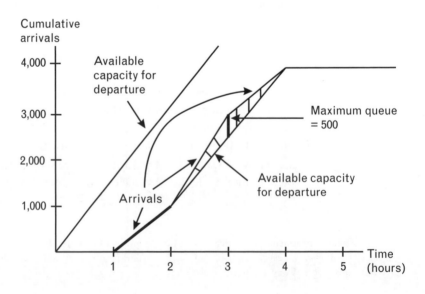

cumulative input and output of this system; on the horizontal axis, it represents time. We know that we can output a maximum of 1,500 vehicles an hour from this system; we do not always output at that rate because the vehicles may simply not be here. So, we have the idea of *available capacity*, which has a slope of 1,500; after one hour, the system *could* output 1,500; after two hours, it *could* output 3,000. In more mathematical terms, this is simply the integral of the constant output function, $\mu(t)$.

Now, consider the accumulation of vehicles on this system. During the first hour, nobody arrives. Between hour 1 and hour 2, we have an input rate of 1,000 per hour, so we have a line with a slope equaling 1,000, going up from zero, and during that hour, as you realize intuitively, everybody who demands service can be provided that service, because the input rate is lower than the capacity of the system.

But, after that, we get into a queuing situation. Between hour 2 and hour 3, 2,000 vehicles arrive. We cannot serve all those vehicles since we are limited by the capacity of the system. So, although 3,000 cars between $t = 0$ and $t = 3$ have demanded service, 2,500 cars will be served over that same period of time: 1,000 from $t = 1$ to $t = 2$ and 1,500 from $t = 2$ to $t = 3$. So at $t = 3$, we still have 500 cars in the system. Between $t = 3$ and $t = 4$, we again have the situation in which capacity is greater than the demand; between $t = 3$ and $t = 4$, we have to service 1,000 newly-arrived vehicles, plus the 500 left over from the previous hour, which we are able to do (exactly in this case) with our capacity of 1,500 vehicles/hour.

Now, the numbers were selected to make this simple; at the end of four hours the system is empty. The queue dissipated exactly at the end of four hours. But for example, suppose vehicles arrive at the rate of 1,250/hour from $t = 3$ to $t = 4$.

You have a queue of 250 cars at $t = 4$.

Yes. How would you represent it on that diagram (see Figure 27.3)? When would the queue be dissipated?

The departure rate is 1,500/hour, so it would dissipate in one sixth of an hour.

This illustrates the concept of deterministic queuing. It is based on the idea of flows and service rates that are deterministic, not

FIGURE 27.3   *Another queuing diagram.*

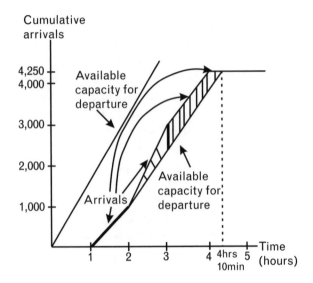

probabilistic, over time. In this formulation, input and output rates *can* be time varying, while still deterministic.

What is the longest queue in this system? Let's go back to Figure 27.2. 500 at t = 3 is the longest queue. What is the longest individual waiting time? Under these assumptions, the longest waiting time is going to be for the car that had the most cars in front of it when it arrived. So, if 500 was the longest queue, and we service cars at the rate of 1,500/hour, the longest waiting time is a third of an hour or 20 minutes.

Computing Total Delay

Another important idea—presented without proof here—is that if one takes the area between the input and output curves—a triangle in this case—the total delay in vehicle-hours is equal to that area. So the area of the shaded triangle equals the total number of vehicle-hours of delay (see Figure 27.4).

Choosing Capacity

Now, going back to some of the early concepts, the question is, "How do you choose the magnitude of $\mu(t)$, assuming that it is a constant

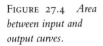

FIGURE 27.4 *Area between input and output curves.*

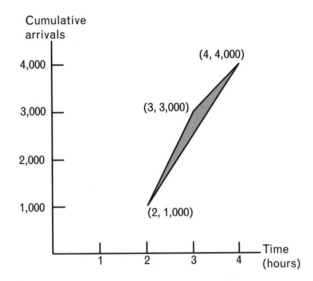

(this *is* a design decision)?" This is fundamentally a capacity versus level-of-service question. One could choose $\mu(t) = 2,000$. In that case the total delay is zero. The system is designed to accommodate the maximum amount of demand.

But capacity costs money and this designing for the peak load can be expensive in systems that have high peak demands. So we usually don't design systems for peak demand. In the example, we split the difference between peak and base load and chose a capacity of 1,500, which introduced some delay.

You can compute total delay in vehicle-hours as a function of the capacity of the system, $\mu(t)$. One could choose 500 as the capacity and one would eventually dissipate all those 4,000 cars. But it would take a long time; it would take 8 hours. So, you have some very long delays. On the other hand, you do not want to overly invest in infrastructure (i.e., capacity) you only need at the peak, for perhaps one hour per day.

A Word on Probabilistic Queuing

Now, remember we have discussed only deterministic flows into our system and deterministic service rates. There are also systems in which vehicles arrive probabilistically and the service time for each is probabilistic as well. This is more complex mathematically and beyond what we will do here. It is worth noting, in passing, that the average waiting time in such systems goes up as the variance in arrival and service times goes up. In general, uncertainty in flows into and out of the system hurt average performance.

A Traffic Light as a Deterministic Queue

Now, here is a special case that is of some interest to us: consider a traffic light as an example of deterministic queuing. Cars are flowing steadily in a traffic stream; they come to a light that is either green or red. If it is green, they go through; if it is red, they stop. For the moment, consider flow in one direction; we will generalize it soon. One can characterize the service of this kind of traffic light system, $\mu(t)$, in the following figure (see Figure 27.5).[2]

During periods of red, there is no output at all—cars are waiting for the red light to turn green. During the periods of green (for simplicity, we will forget yellow lights), there is some output function that describes how quickly cars can get through that light. Now, for the sake of this analysis, we ignore such niceties as the dynamics of a traffic stream starting up, people reacting and lagging. We simply assume a uniform flow. So, this is the inverse of the first example; here the arrival rate is a constant and the service rate changes with time.

FIGURE 27.5 *Service rate and arrival rate at traffic light.*

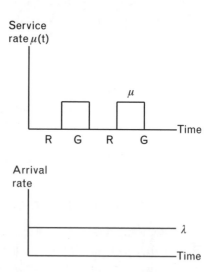

2   Once again, the reader is referred to Mannering, F. and W. Kilaresky, *Principles of Highway Engineering and Traffic Analysis,* 2nd Ed., New York: John Wiley and Sons, 1998, for an excellent and more extensive treatment.

Queuing Diagram for a Traffic Light

Again, we draw the queuing diagram. There is going to be queuing, because some people are going to arrive when the light is red. So we know the arrival rate is a straight line with slope λ. Again, mathematically this is simply the integral of the uniform arrival rate (see Figure 27.6).

Now, plot the departures from this system. Assume that we start with no one waiting in queue. We know that during the red period, nobody departs. What happens when the light turns green? There is a queue built up from the red period. People depart at the rate μ; at some point, the queue is going to be zero. The queue that accumulated during the red period, R, plus the additional cars that arrived while those initial cars were leaving, is emptied. You can see that μ has to be greater than λ because we have to dissipate not only the cars that are in the queue when the light changes, but also those that arrived during that process. Then, once the queue is empty, the cars will simply depart at the steady arrival rate of λ.

Queue Stability

Now, why must the queue be zero before the end of green? What happens if it is not? Let's suppose at the end of the green cycle there are still four cars in the queue. What's going to happen?

The queue will keep growing.

Right. We will end up with four cars left from every cycle. Eventually, under this set of assumptions, it will grow to infinity.

FIGURE 27.6
Queuing diagram at traffic light.

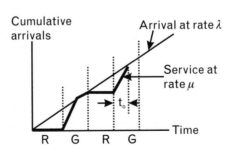

Now, define the time from the light changing from red to green until the queue is dissipated as t_0. A condition for stability—we do *not* have a queue growing to infinity—is that t_0 has to be less than or equal to the amount of green time (G) in this cycle. Putting it mathematically,

If $R + G = C$ (the cycle time),
then $\lambda(R + t_0) = \mu t_0$.

Rearranging $t_0 = \dfrac{\lambda R}{\mu - \lambda}$

If we define $\dfrac{\lambda}{\mu} = \rho$ (the "traffic intensity"),

then $t_0 = \dfrac{\rho R}{1 - \rho}$

For stability $t_0 \leq G = C - R$

When you derive mathematical relationships like this, every once in a while stop and say, "Does this make sense?" For example, let's set $\rho = .5$. The discharge rate from this traffic signal $\mu(t)$ is exactly double the arrival rate. What do you think t_0 ought to be under those circumstances? How long is it going to take to dissipate the queue?

All of the red time.

Right. All the red time is needed. The rate t_0 will be exactly equal to R under those circumstances.

Delay at a Traffic Signal: Considering One Direction

We can again use the idea that the area between the arrival and departure rate curve is the total delay in vehicle-hours. If you go through the geometry of computing what the area is in Figure 27.6, the total delay, D, is as follows:

$$D = \frac{\lambda R^2}{2(1 - \rho)}$$

The total delay per cycle is d

$$d = \frac{D}{\lambda C} = \frac{R^2}{2C(1 - \rho)}$$

What is the maximum delay in this system? Who has to wait the longest in this system? The person who just missed the light has to wait the longest. The maximum delay in the system is simply equal to R, the red time.

Two-Direction Analysis of Traffic Light

The next step in this kind of an analysis is considering two directions. We have to balance the red and green times in various directions. Suppose there are four flow rates: λ_1) Eastbound; λ_2) Westbound; λ_3) Southbound; and λ_4) Northbound (see Figure 27.7). We want to calculate an optimal split between red and green, noting, of course, that red in one direction is green in the other (we are still ignoring yellow).

In choosing a cycle time to minimize delay, measured in vehicle-hours, we have to decide how to split it between red and green as a function of these traffic conditions. Now, to simplify, assume that red time in direction 1 is equal to red time in direction 2. That is, the amount of red time is the same for the flows in the Eastbound and the Westbound directions. One could imagine with left turn signals they could be different, but here we assume $R_1 = R_2$. Also, we assume $R_3 = R_4$.

FIGURE 27.7 *Flows in East-West and North-South directions.*

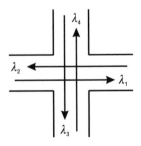

....

Delay at a Traffic Signal: Considering Two Directions

Further assume that the flow capacity—Eastbound, Westbound, North-bound, Southbound—is the same. Again, that is not necessarily the case. You could certainly have different numbers of lanes in different directions. We now can compute D_1, the delay in direction 1 (Westbound).

$$D_1 = \frac{\lambda_1 R_1^{\,2}}{2(1-\rho_1)} \text{ where } \rho_1 = \frac{\lambda_1}{\mu}$$

We can write similar expressions for D_2, D_3, and D_4. We want to minimize D_T, the total delay, where

$$D_T = D_1 + D_2 + D_3 + D_4$$

Choosing an Optimum

Remembering that

$$R_2 = R_1$$
$$R_4 = R_3 = (C - R_1)$$

we want to minimize D_T where

$$D_T = \frac{\lambda_1 R_1^2}{2(1-\rho_1)} + \frac{\lambda_2 R_1^2}{2(1-\rho_2)} + \frac{\lambda_3 (C-R_1)^2}{2(1-\rho_3)} + \frac{\lambda_4 (C-R_1)^2}{2(1-\rho_4)}$$

To obtain the optimal R_1, we differentiate the expression for total delay with respect to R_1 (the only unknown) and set that equal to zero.

$$\frac{dD_T}{dR_1} = \frac{\lambda_1 R_1}{1-\rho_1} + \frac{\lambda_2 R_1}{1-\rho_2} - \frac{\lambda_3 (C-R_1)}{1-\rho_3} - \frac{\lambda_4 (C-R_1)}{1-\rho_4} = 0$$

Let's try a special case; assume the flows in all four directions are the same:

$$\lambda_1 = \lambda_2 = \lambda_3 = \lambda_4$$

Therefore, $\rho_1 = \rho_2 = \rho_3 = \rho_4$
The result, then, is

$$R_1 = \frac{C}{2}, \quad R_3 = \frac{C}{2}$$

This makes sense. If the flows are equal, we would expect the optimal design choice is to split the cycle in half in the two directions.

Now, we generalize. Assume that

$$\lambda_1 = \lambda_2 \qquad \lambda_3 = \lambda_4 = k\lambda_1$$

Differentiating the expression for total delay D_T and setting it equal to zero, we get

$$\frac{\lambda_1 R_1}{1-\rho_1} + \frac{\lambda_1 R_1}{1-\rho_1} - \frac{k\lambda_1(C-R_1)}{1-k\rho_1} - \frac{k\lambda_1(C-R_1)}{1-k\rho_1} = 0$$

Eventually, we obtain

$$R_1 = \frac{Ck(1-\rho_1)}{(1+k-2k\rho_1)}$$

as the expression for the optimal setting.
The expression

$$\frac{k(1-\rho_1)}{(1+k-2k\rho_1)}$$

is the fraction of the cycle time in the North-South direction.

Some Special Cases

It is good practice to make sure the expression works for special cases for which we know the answer.

$$\text{So, if } k = 0, \text{ then } R_1 = 0$$

This makes sense. If there is zero flow E-W, we give all the green to N-S.

If $k = 1$, we have equal flows in all directions which is the case we just considered:

$$R_1 = \frac{C}{2}$$

which is the result we obtained before.

Let's try another more interesting case. Suppose

$$k = \frac{1}{2}$$

That means 2/3 of the flow is North-South and 1/3 is East-West. We obtain

$$R_1 = \frac{C(.5)(1 - \rho_1)}{(1.5 - \rho_1)}$$

What happens in light traffic? Compute R_1 as ρ_1 approaches zero.
R_1 approaches $\frac{C}{3}$ as ρ_1 approaches zero.

Now, let's see what happens as traffic grows. Try $\rho_1 = .25$.
Then $R_1 = .3C$ $G_1 = .7C$
Now we need to check that the queues in both directions can be emptied in the available green time. Remember we earlier had an expression for t_0, the time needed to clear the queue.

$$t_0 = \frac{\rho_1 R_1}{1 - \rho_1} \text{ in the North } - \text{South direction.}$$

$$t_0 = \frac{.25(.3C)}{.75} = .1C$$

So, we have time to clear the queue since $t_0 < .7C$, which is the green time North-South.

Now, we have to check the East-West direction as well.

$$\rho_3 = k\rho_1 = .5(.25) = .125$$

$$t_0 = \frac{\rho_3 R_3}{1 - \rho_3} \text{ in the East } - \text{West direction}$$

$$= \frac{.125(.7C)}{.875} = .1C$$

Green in the East-West direction is .3C, so the queue clears. We are not surprised since the traffic intensity was .25 in the heavy (North-South) direction and only .125 in the light (East-West) direction.

Let's examine what happens as traffic intensity grows.

$$\rho_1 = .5$$

$$R_1 = .25C \qquad G_1 = .75C$$

$$t_0 = .25C \text{ in the North } - \text{South direction}$$

Again we have capacity North-South because $t_0 < G$ in that direction. Now check the East-West direction.

$$\rho_3 = k\rho_1 = .5(.5) = .25$$

$$R_3 = .75C \qquad G_3 = .25C$$

$$t_0 = \frac{.25(.75)C}{(.75)} = .25C$$

This is just enough capacity. The variable t_0 is exactly equal to the green time East-West.

Let's now try $\rho_1 = \dfrac{2}{3}$

Flow is now heavy relative to the capacity in the North-South direction

$$R_1 = .2C \qquad G_1 = .8C$$

for an "optimal" split between North-South and East-West. In the North-South direction

$$t_0 = \frac{\left(\dfrac{2}{3}\right)(.2C)}{\left(\dfrac{1}{3}\right)} = .4C$$

which is less than .8C. So, the North-South queue clears. However, in the East-West direction

$$R_3 = .8C \qquad G_3 = .2C$$

$$t_0 = \frac{\left(\dfrac{1}{3}\right)(.8C)}{\left(\dfrac{2}{3}\right)} = .4C$$

But G_1 is only $.2C$. There is not enough green in the East-West direction to clear the queue. So we need to share the splits to assure the East-West queue does not grow without bound. In this case,

$$G_3 = .4C \qquad R_3 = .6C$$
$$G_1 = .6C \qquad R_1 = .4C$$

Remember that t_0 in the North-South direction was $.4C$, but $G = .6C$, so we are feasible.

Now there is a level of ρ_1 at which we cannot clear the queue (in one direction or the other).

A good exercise is to find that value of ρ_1 (remembering that in this formulation $\rho_3 = k\rho_1$).

The deterministic queuing model idea is useful in many transportation applications, if the deterministic assumption is appropriate. In this case, we know the arrival and service rates are really stochastic, but have made a judgment that in our analysis we can treat the system as deterministic—you can wring a lot of insight about system performance from a deterministic queuing model.

CHAPTER 28

Urban Public Transportation

Urban Public Transportation

Next we consider urban public transportation in a continuation of our treatment of traveler transportation. We will focus on United States public transportation, although many of the fundamental concepts we will discuss will have international applications.

LOS Variables for Urban Travelers

To begin, we review our level-of-service variables for travelers. Travel time, reliability, cost, waiting time, comfort, safety, and security are all examples of level-of-service variables that are relevant to any traveler transportation mode. We have characterized the automobile mode, with which public transportation is in direct competition, as very convenient. One rides in the comfort of a climate-controlled automobile listening to the stereo; the concept of waiting time for service is nonexistent—one goes when one wants to go. The cost of using an automobile may even be low, especially if one does not consider the capital cost of the vehicle, as people have a tendency not to—"I have to own the automobile anyway."

How Public Transportation Measures Up

Public transportation tends to falter on several of these dimensions. For example, comfort in a crowded rush-hour subway car is not high; one has to wait for service depending on the service frequency of the mode one is considering. Travel time may be greater or less than that of an automobile, depending on the circumstances.

A hard-to-quantify level-of-service variable is self-image—how one feels about oneself as a function of the mode you use. In much of Western society there is a very positive self-image associated with driving and a much less positive self-image associated with taking public transportation.

Security is another concern. For example, crime in the New York City subways gets a very high profile in the *New York Times*, although recent rashes of car-jacking make automobile transportation subject to some security concerns as well.

Availability of service is a question; some urban subway systems shut down during the late night/early morning hours. The Paris Metro does exactly that. Many urban bus services do not run during late evening hours. This makes the lives of night workers very difficult if they need to depend on public transportation.

Safety is another level-of-service variable. Again, while auto, judging by aggregate statistics, is a less safe mode than transit, people tend to discount that statistical difference, feeling that they, in control of their own vehicle, can drive safely. Travelers tend to focus on the occasional major transit accidents as evidence of a lack of safety in public transportation modes.

Accessibility to Service

Of particular importance is the question of accessibility to service. Earlier, we considered the Federal Transit Administration's (FTA) perspective on their difficulty in serving the very spread-out land-use patterns developed in suburban America. This is a critical concern. With urban sprawl, population densities are such that viable public transportation, which depends on consolidating the demands of many users to provide economies-of-scale, is virtually unachievable. So the development of public transportation is very much impeded by the urban structure that has developed in the United States in the post-World War II era.

Accessibility to public transportation service takes on another dimension, if we consider an intermodal car-public transportation journey. Commuters who would like to park-and-ride are constrained by parking availability in parking lots at stations. Drivers complain that if they arrive at the parking lot after 8:00 a.m., the parking lot is full. An option is satellite bus services that would circulate through the suburban neighborhoods, picking up people and bringing them to the suburban stations by van, as a mechanism for overcoming the parking constraint.

This illustrates the systemic nature of transportation. One has to consider more than the simple line-haul characteristics of, say, the rail system, which in itself may provide excellent service. To think about one's customers' needs more broadly, one has to recognize that the customer has to get to the system. And in this instance, limitations on parking lot size is preventing that from happening.

Types of Urban Public Transportation Service

Let's talk about the various services that we include in urban public transportation.

Conventional Bus

We start with the conventional bus. These services typically operate on fixed service routes, with stops specified. A typical urban bus would have a total capacity of about 60 passengers. Technologies would include internal combustion engines, diesel engines, and what are called trolley buses—buses that run on electric power.

Para-Transit

A second service is para-transit. Para-transit is a catch-all phrase for nonconventional services, often using smaller or specially-equipped vehicles. For example, para-transit services would include special services for the elderly and handicapped. These services would typically pick up and deliver people from their home to some destination like an elder-care facility.

In some cities, there are less formal modes of transportation that can be characterized as para-transit. For example, in San Juan, Puerto Rico, there is a system of públicos. These are van-type vehicles, often operated by an individual entrepreneur, which do not go on a fixed route but rather pick up passengers and deliver them to their destination, informally providing ride-sharing among groups of passengers with disparate origins and destinations.

Públicos often provide "one-to-many" service. A público could wait outside of a transit station and pick up travelers heading for the same general area of the city. The público driver would then travel in some efficient manner to provide a reasonable travel time at a reasonable cost. In some cities, these services are called "jitneys." Sometimes they run on semi-fixed routes, deviating from the routes occasionally. Often, they are totally free-form.

Demand-Responsive Service

There are demand-responsive urban transportation systems. A traveler would call for service, perhaps from his or her home or office, indicating the origin and destination, and a dispatcher, perhaps assisted by a computerized system, would route a vehicle dynamically to pick up and deliver the passengers in an effective manner. This is ride-sharing; the hope is that these services achieve some economies-of-scale through shared capacity while providing a "taxi-like" level-of-service. Indeed, we can consider conventional taxi service as a public transportation service.

Rail Systems

So far, we have introduced highway-oriented public transportation modes. Fixed-rail systems are important as well.

Subways

Large urban areas often have rail systems categorized into "heavy rail" and "light rail." The MBTA here in Boston, the MTA in New York City, the Washington Metro, and the CTA in Chicago are examples of major subway systems in the United States. These heavy rail systems, which are typically a combination of underground, grade-separated, and above-ground infrastructure, have the capability of transporting large numbers of people very efficiently along fixed corridors. Light systems often operate at grade and are lower-capacity and lower cost compared with heavy rail.

Some of our cities are virtually inconceivable without urban rail. New York City has been brought to its knees by subway strikes. Other cities like Los Angeles, a more automobile-oriented city, has functioned without a rail system for many years. Today the Los Angeles subway system is marginal, serving a very modest area of a quite spread-out metropolitan region.

It is worth noting that if one looks at total public transportation ridership in the United States, New York City represents half that total ridership. Talk about industry concentration! Half of public transportation ridership in the United States is in one urban area, in which perhaps 7% of the nation's population resides.

Commuter Rail

Also, we have commuter rail. These are typically trains from suburban areas into downtown. Commuter rail typically extends further into the suburbs than do urban metro systems. Cities like New York and Chicago are particularly dependent on these commuter rail services, with people often commuting distances in excess of forty miles to work downtown in these cities.

Intermodal Services

Another service type is intermodal, the conceptual equivalent of freight intermodalism. It has real relevance in the urban traveler domain. We have bus services serving as feeders to fixed rail systems. We have already introduced park-and-ride, with people driving to public transportation, parking their car, and taking the public transportation system into the urban area. There are any number of opportunities here. The key, as it was in freight, is proper coordination among the modes. The idea of intermodalism is to use the inherent advantages of various modes, but those inherent advantages can be quickly dissipated if the transfer between modes is poor. So coordination and careful scheduling is critical to success in intermodalism.

Public transportation has tried to redefine itself in recent years. Brian Clymer, who was the FTA administrator in the Bush administration (1988-1992), took the strong position that public transportation is everything that is *not* single-occupancy vehicle (SOV) automobile transportation. So he considered carpools, van pools, and the development of HOV lanes as an appropriate part of the transit domain. In our car- and highway-oriented society, that is a pragmatic position for a public transit administrator to take. Operate within the highway-dominated system as best one can to try and reduce SOV transportation.

Public Transportation Patronage

Figure 28.1 shows the trends in public transportation patronage since the beginning of the 20th century.

FIGURE 28.1 *Public transportation patronage in the United States (Source: APTA 1999 Transit Fact Book, American Public Transit Association, Washington, D.C., 1987).*

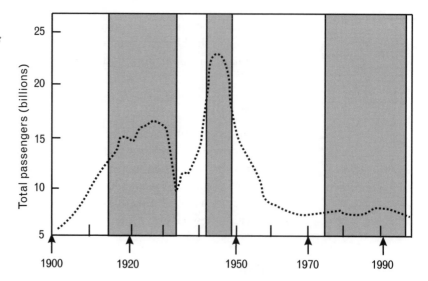

At the start of the twentieth century, passenger patronage was about 6 billion riders per year. To set this in time, the MBTA in Boston had been in operation for several years, as was the New York City subway system. Various kinds of streetcars and horse-drawn vehicles were part of the public transit fleet at that time.

In the next quarter century until 1925, the transit industry grew steadily to a peak of 18 billion passengers annually. This was a period of substantial urban growth in the United States. It represents a time in United States history when the nation shifted from an agrarian to a manufacturing economic base. People by the millions migrated from farms into urban areas for better paying manufacturing jobs. At the same time, automobile ownership was for only the richest in society. Public transportation was the only option for large numbers of people in urban areas.

Beginning in the late 1920s and into the 1930s, transit ridership dropped off precipitously, reflecting an economic depression of unprecedented proportions. People had no jobs; therefore there was no journey to work, and transit ridership reflected that. As the country came out of the depression in the mid-1930s, transit ridership grew again. And with United States participation in World War II, transit ridership grew substantially to a peak of 24 billion in 1945. Jobs were plentiful on the homefront and automobiles were again virtually unavailable because war-time manufacturing had turned to tanks and planes, not private automobiles. Tight fuel rationing held down automobile use as well.

However, with the end of World War II, pent-up economic demand, explicit government policies to develop the highway system, and the desire of many for a single-family home in the suburbs led to the urban land-use patterns that we discussed earlier. The 24 billion passengers carried in 1945 turned out to be the high watermark for public transportation in the twentieth century. Public transit use fell off precipitously beginning at that time and currently is at a level of about 10 billion passengers per year. This is modestly up from the bottom of the mid-1980s, but not significantly. So the current transit industry in the United States is about 40% the size of its post-war, 1945 peak, and unlikely, in the current environment, to grow significantly.

Importance of Bus Services

On a national scale, about two-thirds of the current transit patronage is carried by buses. For those of us who are large-metropolis- and rail-oriented, this comes as somewhat of a surprise. But when you think about it, the number of cities with subway systems is not large. Virtually all public transportation in cities of less than several million is by fixed-route bus and, even in cities like New York, a substantial fraction of riders go by bus.

Temporal Peaking and Its Implications

Peaking is of particular importance in urban public transportation. There is tremendous peaking in demand during the morning and after-noon rush hours, more so than the peaks that occur on the highway network. Transit ridership is dominated by the journey-to-work which causes this peaking phenomenon. While frequent service is the rule during the peak hour, vehicles are often very crowded, and passengers are not very comfortable.

Now in any kind of service operation, during the period of peak demand you require peak staffing. The problem in the transit industry is that those peaks occur on the order of eight hours apart. To have an adequate number of drivers on hand for both peaks would suggest that people would work a "split-shift." Simply put, this means they would work in the morning, go home during the middle of the day when the demand is low, and return for the afternoon peak. Of course, people tend not to like those kinds of shifts. For that and other reasons, absen-teeism is a chronic problem on transit properties across the country.

Simply put, these kinds of jobs are not viewed as particularly desirable, given the stress of operating in rush-hour conditions, and the relatively low pay and status.

So, labor productivity is one of the key issues that transit managers face. And one could argue that this is a direct result of the strong peaking in the demand for public transportation.

Characteristics of the Public Transportation Industry: A Personal View

Let us proceed now to characterize the industry, recognizing this is the personal (and possibly biased) view of the author.

The public transportation industry is:

1. Financially marginal. The industry is heavily subsidized. The MBTA pays only about one-third of its operating costs from the farebox. This does not include investment costs for new vehicles or infrastructure. While all transportation is subsidized to some extent, the United States transit industry is more heavily subsidized than most other modes. This leads to a chronic under-investment in physical plant and chronic deferred maintenance of both vehicles and infrastructure.

2. Not very innovative. Adoption of new technologies by public transportation agencies is not aggressive. Now perhaps—maybe almost certainly—this can be attributed to their financial condition. But, whatever the reason, new technology tends not to be adopted very quickly.

3. Public-sector dominated. While there have been some recent attempts to privatize public transportation, particularly bus services, for the most part public transportation continues to be operated by government. In this milieu, salaries tend to be lower, and it is more difficult to attract first-rate people to manage these properties.

4. Subject to political pressure. As part of the public sector, there are political pressures on the industry that are substantial. Decisions on a variety of matters—operations, staffing—are often made on a political rather than a service and cost basis.

5. Subject to poor labor relations. These are endemic to the industry. Absenteeism in the public transportation industry is among the highest of any industry in the United States, public or private.

6. The victim of a public image problem. Public transportation is viewed as a low-service-quality, low-efficiency industry, taken usually only by captive riders—people who have no other option, such as automobile. Few boast about taking public transportation to work (outside of the academic world, of course).

7. Not especially active in market development. Marketing in public transportation is currently more a research area than it is an active part of management. Recently, market studies have been undertaken, aimed at identifying what travelers really think about public transportation and how services could be designed that would be attractive to them [1]. This is not a new idea in American business, but it has not been developed in any major way in public transportation organizations. Again, one could argue that this all derives from the financial woes of the industry.

8. Marginalized by low-density land-use patterns in United States cities. This dominant issue may well prove intractable.

The picture is bleak. While new technologies are helping many transportation industries, their deployment in public transportation is modest. Land-use patterns have developed that make public transit a less-than-useful mode in many urban areas. Political pressures are critical and the industry is chronically highly-subsidized and underfunded.

In this environment, it is a brave manager indeed who can push towards efficiency and effectiveness, but there are some out there in American industry. First-rank public transportation organizations exist in Houston, Texas; Portland, Oregon; Seattle, Washington, among others; and New York City, for all its woes, transports an extraordinary number of people to and from their jobs and other origins and destinations every day—perhaps not in the greatest comfort but from a throughput point of view, it is difficult to fault.

Putting aside this bleak picture, we now take a look at some modeling approaches to understanding some of these issues. (These approaches can be used for other modes as well.)

....
Life-Cycle Costing

We have talked about the financial woes of transit causing under-investment in infrastructure and vehicles and under-maintenance of vehicles and infrastructure as well. Let us use the idea of life-cycle costing to understand these issues. Figure 28.2 shows a flow of costs, both capital and maintenance, over time for a transportation infrastructure facility.

We invest at the beginning of a project with capital and then maintain it over a period of years. Here we show a flow of maintenance costs designated by M_n, which may vary from year to year. There is always a trade-off between capital and maintenance costs. High capital costs often lead to lower maintenance costs and vice versa.

In life-cycle costing, we use the concept of discounted cash flow, which incorporates the time value of money into an overall cost for a property.

$$DCF = C_C + \sum_{n=1}^{\infty} \frac{M_n}{(1+i)^n}$$

where C_C is the capital cost and i is an interest rate or "discount rate" that reflects the time value of money as well as a risk factor.

Keeping the discussion simple, we will not worry about the flow of "benefits" or "revenues" in this discussion.

Maintenance costs can be represented by the following equation:

Maintenance Costs = $f($ Quality of initial construction,

current state of infrastructure,

wear-and-tear caused by traffic)

Think of infrastructure as shown in Figure 28.3.

FIGURE 28.2
Life-cycle costs.

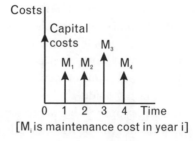

[M_i is maintenance cost in year i]

FIGURE 28.3
Infrastructure quality.

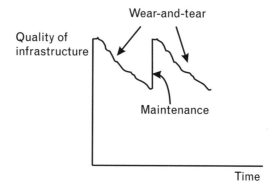

Wear-and-Tear and Maintenance

The quality of the infrastructure at any time is degraded by accumulated wear-and-tear caused by use—traffic—and improved by maintenance. So we plot quality of infrastructure as a function of time, as shown in Figure 28.3.

There are several factors in considering a maintenance strategy. First, the value of a dollar of maintenance expense is a function of the current quality of the infrastructure. If you have a well-maintained facility—it has a high dollar value—the benefits of one dollar of maintenance are higher than if one has a poorly maintained facility. This is shown in Figure 28.4. The benefit of $M of maintenance would slowly decrease with time, as shown in Figure 28.5.

The Problem with Deferred Maintenance

So delaying—or deferring—maintenance, which is a strategy often used in the stringent economic times characteristic of the transit industry, is a problem. If we delay maintenance, the "value" of the infrastructure goes down and the benefit of a maintenance dollar goes down as well.

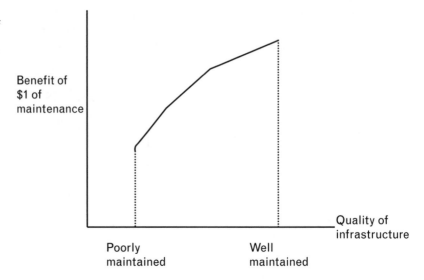

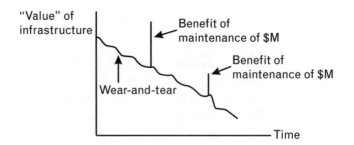

However, it is even worse than that. There is empirical evidence to show that the amount of wear-and-tear caused by a unit amount of traffic increases as the "quality" or "value" of the infrastructure decreases. For example, for a fixed rail system, if you have a right-of-way in poor shape, the amount of wear-and-tear caused by a single train is greater than if you have infrastructure in good shape subject to that same train. (A right-of-way in poor shape also causes more wear-and-tear on the vehicle, also, but that is another story). This is illustrated in Figure 28.6.

So, if we delay or skimp on maintenance, the quality of our infrastructure goes down more quickly. We buy less for our maintenance dollar. Further, traffic causes more damage. More wear-and-tear is caused by a fixed amount of traffic.

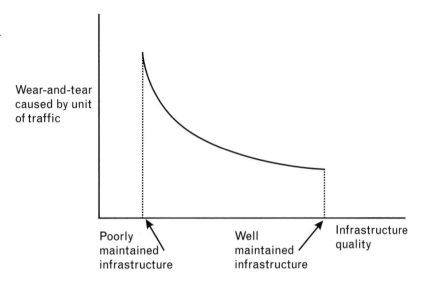

FIGURE 28.6   *Wear-and-tear as a function of infrastructure quality.*

Quality of Infrastructure and LOS

But, it gets still worse. There is a relationship between the level-of-service as observed by customers and the quality of the infrastructure. As quality of infrastructure deteriorates, trains go more slowly—perhaps they derail—then the level-of-service deteriorates, as shown in Figure 28.7.

So, as the quality of our infrastructure deteriorates, level-of-service deteriorates, and considering the equilibrium framework, traffic volumes will deteriorate. So revenue goes down and there are even fewer dollars to spend to improve our infrastructure by counterbalancing the effects of wear-and-tear through maintenance.

FIGURE 28.7   *LOS as a function of quality of infrastructure.*

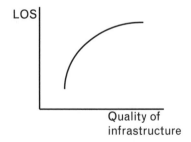

The Vicious Cycle

This is often and appropriately called "the vicious cycle." Once things start to get bad, it is very expensive to make them right again. This is true in many different kinds of systems. It has particular impact in the transit industry where money is always so chronically short and political pressures often dominate reasonable management decisions.

Service Design

We consider service design for public transportation. Service design includes routes, vehicles, frequency of service, hours of service, level and structure of fares, and supporting services, such as parking and station configuration. Service design will differ depending upon the technology being used. Clearly if one has a fixed rail system, the ability to change routes is much more limited than it is for an urban bus system.

Fare systems vary around the industry. For generations, the New York City subway system has had a flat fare system, with one fare regardless of distance traveled. Other systems, like the Washington Metro, have a fare structure in which one pays as a function of the distance traveled and also the time of day. More advanced technology is increasingly in use in transit services for collecting fares. The Washington Metro has had a fare card for many years, as has BART in San Francisco. And New York is deploying that technology as well. This enables fare structures to be more sophisticated than the simple flat fare system, and can include bulk discounts for frequent users.

An important consideration in fares is the extent to which they encourage intermodality. For many years, New York City had its infamous 2-fare zones in which travelers paid a fare to ride the bus, which took them to the subway, so they could then pay an additional fare for the subway ride. Many systems encourage intermodality by having an integrated fare system where one pays no additional fare or a modest additional fare for use of the second mode.

Network Structure

A very important service design question is that of network structure. In their excellent text, Gray and Hoel [2] discuss a number of different network structures for public transportation, including radial patterns, grid network, radial criss-cross, and trunklines with feeders.

The Vehicle Cycle

The vehicle cycle is a fundamental design element of transportation systems and public transportation is no exception. An important problem in transportation systems design is the sizing of a fleet—how many vehicles do we need to supply service at some service frequency on a particular route? Frequency is expressed in terms of the number of vehicles per hour on a particular route. From the point of view of the traveler we speak of headway, which is the time interval between consecutive vehicles. Frequency and headway are of course the inverse of one another.

The basic equation for sizing a fleet is as follows:

$$NVEH = \frac{VC}{HEADWAY}$$

where $NVEH$ is number of vehicles in the fleet; VC is the vehicle cycle on this route—the time it takes the vehicle to traverse the entire route; and $HEADWAY$ is the scheduled time between consecutive vehicles.

Alternatively,

$$NVEH = FREQUENCY \cdot VC$$

where $FREQUENCY$ is the number of vehicles per unit of time passing a point on the route. Note this is identical to the calculation of the number of freight cars in the fleet in Chapter 16's section on Fleet Size Calculation.

A Simple Example

Suppose we have a route that goes from east to west and then returns from west to east, with the travel time being one hour in each direction, as shown in Figure 28.8. VC in this case is two hours.

Suppose we want a service frequency of four vehicles per hour. From the passenger's—our customer's—viewpoint, this means a

FIGURE 28.8
Vehicle cycle.

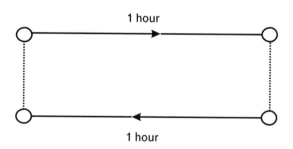

headway of 15 minutes or .25 hours. Using the above equation, the required number of vehicles would be eight.

How can you reduce the number of vehicles needed in this service? First, we could have less frequent buses running, say, at a headway of .5 hours rather than .25 hours. The number of vehicles required goes down to four.

Another method could be to shorten the vehicle cycle. Suppose we had fewer stops along the route and the vehicle cycle dropped from a total of 2 hours to 1.5 hours. In this case, the number of vehicles we would require would go down to six.

Of course, it is not as simple as all this. Given stochasticity and the propensity of such systems to get out of balance, we observe some interesting behavior in these systems.

Bus-Bunching: An Explanation

For example, one phenomena that has been much commented upon is bus-bunching. The traveler waits a long time for a bus to arrive and then the buses come in a convoy of two or three. Think about why this might happen.

Let us assume that people arrive uniformly over time at a bus stop. If the transportation service is frequent enough, people will not try to "catch" a particular bus, but will arrive uniformly. Figure 28.9 shows the uniform build-up of people at a particular bus stop.

We assume that people begin to arrive as soon as the previous bus departs and the number of people waiting at the stop builds up linearly over time. When the bus arrives, people depart from the bus and the people at the bus stop get on the bus. The amount of time the bus spends at the stop—the dwell-time—is proportional to the number of people getting on and off.

FIGURE 28.9 *Why buses bunch.*

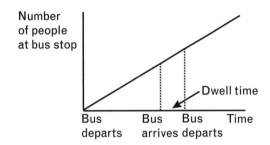

Now let us suppose that a bus departs the terminal a few minutes earlier than it should have according to the schedule, which is designed to achieve uniform headways. What will happen now is that the interval between that bus and the following bus will be greater than normal. There will be more people waiting for the second bus at each of the bus stops. The dwell-time at each of the bus stops will be longer. The third bus departs according to schedule, but the second bus is slowed by excessive dwell-times—remember during longer dwell-times still more people arrive—and the third bus soon catches up to the second bus, leading to the bus convoy or bus-bunching idea.

The Vicious Cycle: Another Example

This is another example of the vicious cycle that we encounter in systems with feedback. We used this idea in describing the relationship between infrastructure maintenance in public transportation and level-of-service. The same kind of phenomena operates here. The more out of balance a system gets, the stronger the forces are to force it still further out of balance, with service deteriorating.

Needless to say, bus drivers who chronically leave the terminal early are not well-liked by their peers, who pay the price by arriving late at the opposite terminal and having a very crowded bus along the route to boot. So there is some peer pressure to retain schedules and proper spacing.

Control Strategies

Let us now discuss some strategies that transit properties—here using rail as an example—can use to improve operations. We call these *control strategies*.

Holding Trains

This strategy at stations is a mechanism for making headways more uniform (the previous example illustrated why uniform headways is an important characteristic of such systems). Holding trains is easy to implement and is more common than the strategies described below.

Station Skipping

Under this strategy a train will skip a particular station as a mechanism to make headways more uniform and to improve the overall throughput of the system. This is shown in Figure 28.10. Station-skipping illustrates the fact that individual passengers gain and lose as a result of various operating strategies. If you were one of the passengers that wanted to get off at the station that was skipped, you would have to get off the train at a previous stop and wait for another train that was not going to skip that station. On the other hand, passengers who were not intending to get off at that station will have a better trip. And the system as a whole will operate closer to optimum.

Short Turning

In this strategy a vehicle will not go all the way to the east-most terminal, in Figure 28.11, but will rather stop short of that terminal and come back the other way. Again, this would be used as a mechanism to

FIGURE 28.10
Station-skipping.

FIGURE 28.11
Short-turning.

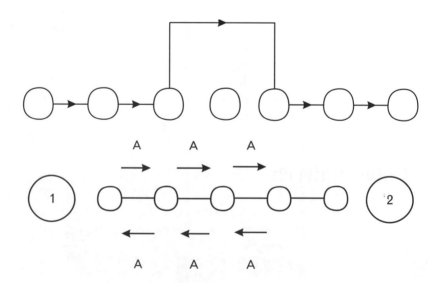

improve the overall throughput and vehicle spacing on the route. "A" is a short-turning train, which returns to "1," without reaching the east terminus "2."

As with station-skipping, short-turning may be inconvenient for particular passengers. If you are traveling in an easterly direction and want to go to the end of the line, you have to get off and wait for the next train to take you there. On the other hand, the operators are presumably optimizing the system for passengers on the whole. Of course, in fixed rail systems we can only turn where it is physically possible to do so, given the track layout.

Need for Real-time Information

All of these approaches suggest a need for real-time information. The operator needs to know where the trains are. We would like to know the passenger loadings on trains and the number of people waiting at particular stations in order to make the above kinds of decisions in a reasonable way.

ITS—Public Transportation Applications

The ITS concept, described in Chapter 24, can be applied to public transportation. These applications, known collectively as Advanced Public Transportation Systems (APTS), include such technologies as automatic vehicle location and automatic passenger counters, which can provide the basis for more efficient fleet management systems, both in fixed rail and bus systems [3].

Intermodal Transfers

Another important ITS application is technologies that expedite transfers between vehicles (be they of the same or different modes). As with freight systems, transfer points often cause inefficiencies and service problems. ITS has the potential for providing the management control so that schedules are coordinated, as well as the traveler information to expedite transfers and make service better.

Traveler Information Through ITS

Traveler information is important—the idea here is to provide user kiosks or information at home or work that would give real-time information about actual vehicle schedules. This is information that would be updated frequently to allow passengers to know exactly what they could expect and also help them with routings through complex transit networks.

Tying these thoughts together, the overall notion is that there are operating strategies that would allow transit systems to operate more effectively. These strategies are information-driven and ITS technologies can be a boon to the transit industry both in improving operations and service and in providing timely information to travelers. The latter in and of itself could be an important market initiative for the public transportation industry.

The ITS community is certainly hopeful that the transit industry will use some of these ideas, and some of the more innovative properties have already begun to do so. However, given the financial stresses on the transit industry nationwide, there is some pessimism about how quickly they can take advantage of these new technologies.

Fares, Ridership, and Finance

Let us now consider the relationship between setting of fares in transit and the overall financial situation that the transit property faces. The overall relationships are shown in Figure 28.12. Revenues are simply the product of the fare multiplied by the volume in riders (assuming a flat fare system). In this analysis, we assume that costs are not a function of volume to a first approximation. That is, the trains or buses operate pretty much the same, independent of volume carried, for "small" changes in volume.

The financial situation of the transit property is governed by "profit," which is simply equal to the total revenues minus the total operating cost. This suggests certain strategies. One could, for example, cut fares and raise volume—the number of riders. Depending upon the form of the demand curve—the function that relates ridership to fare—it is possible that cutting fares may lead to a better financial situation for the transit property.

FIGURE 28.12 *Fares,*
ridership, and finance.

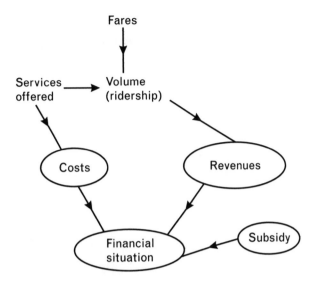

Various Demand Functions

Consider several different hypothetical demand functions.

Linear Demand

We begin with the linear function shown in Figure 28.13. The
equation for this straight line can be written as follows.

$$V = -\left[\frac{V_0}{F_{MAX}}\right] \cdot F + V_0$$

Define R as revenue.

FIGURE 28.13 *Linear*
demand function.

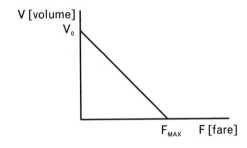

$$R = F \cdot V = F\left[\frac{-V_0}{F_{MAX}}\right] \cdot F + V_0 F$$

Choose F such that R is a maximum.

$$\frac{dR}{dF} = 2F\left[\frac{-V_0}{F_{MAX}}\right] + V_0 = 0$$

So, the level of fare that optimizes revenue F_{OPT} is as shown below:

$$F_{OPT} = \frac{F_{MAX}}{2}$$

Obviously the optimal level of fare depends upon the functional form of the demand relationship.

Parabolic Demand Curve

If we chose a parabolic demand curve, as shown in Figure 28.14, we would have a different optimal level of fare.

In mathematical terms:

$$V = k(F - F_{MAX})^2$$
$$if\ F = 0, \quad V = V_0$$
$$V_0 = kF_{MAX}^2$$
$$k = \frac{V_0}{F_{MAX}^2}$$
$$V = \frac{V_0}{F_{MAX}^2}\left[F - F_{MAX}\right]^2 ; 0 \le F \le F_{MAX}$$
$$\frac{dV}{dF} = 0\ \text{when}\ F = F_{MAX}$$

FIGURE 28.14
Parabolic demand function.

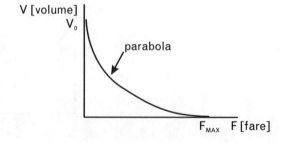

Recalling that R is revenue,

$$R = F \cdot V = F \left[\frac{V_0}{F_{MAX}^2} \right] \left[F - F_{MAX} \right]^2$$

$$\frac{dR}{dF} = \frac{V_0}{F_{MAX}^2} \left\{ \left[F - F_{MAX} \right]^2 + 2F \left[F - F_{MAX} \right] \right\}$$

$$\text{For } \frac{dR}{dF} = 0$$

$$F_{OPT} = \frac{F_{MAX}}{3}$$

"Real" Demand Function

What do you think the demand function really looks like? Many in transit management believe it looks as shown in Figure 28.15.

The horizontal line would reflect little or no change in demand (inelastic demand) as a function of fare for some range of fares. So why not simply raise fares and hence revenue?

Equity

Even if management is right about inelastic demand and the fare increase leads to greater revenues because people are captive, the equity issues remain. Who is being negatively affected by a fare rise? Often it is the most disadvantaged people in society.

If management is wrong, and as a result of a fare rise volume goes down, as shown in the lower right-hand branch of Figure 28.15, revenues go down, exacerbating the system's financial condition. There may be broader macro-economic impacts on the region as

FIGURE 28.15
*Inelastic demand
function.*

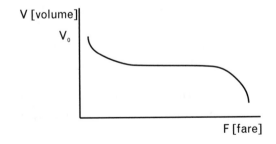

people find it more difficult to get to work and perhaps do not go to work as a result.

Air Quality

There will be air quality issues if more driving occurs. There are safety issues as people shift to the less-safe auto mode from the more-safe transit mode.

Vicious Cycle

Further, if revenue goes down and we under-maintain, the vicious cycle when service quality goes down as well has already been introduced. Certainly the issues in this kind of analysis are subtle. Making predictions in the complex socio-political milieu that we face is difficult indeed. Modeling can help us understand these issues but, as always, understand that modeling is just part of the answer. The results of models must be tempered with professional judgment, recognition of political reality and even some good old common sense if we are to come up with a practical solution or decision in a complicated situation of the sort that we deal with. The fact that we have a sophisticated model that produces an answer does not necessarily mean we have a usable answer for the real world.

Some Other Approaches

We conclude our urban public transportation discussion by noting some approaches that experts in the industry have advanced as a way to improve public transportation and, we would hope, the lot of urban America more generally speaking.

The idea of dedicated busways has been applied. This is basically highway infrastructure that is limited to public transportation vehicles—in this case, buses. The politics here can be difficult in that SOV drivers will note the "under-utilized" infrastructure. Whether it is underutilized on a travelers-per-hour basis is less clear—and that is the point.

David Jones, a transit expert, talks about the need for structural change for transit [4]. He has a recipe for improvement of the industry.

He includes development of new more flexible fare structures, and a market-oriented organizational structure—an area in which transit has badly lagged. Jones speaks of the need for a different mix of vehicles, ranging from large to small, and the provision of a wider diversity of service offerings. He does not feel that the "one-size-fits-all" ethic of the transit industry works in modern society. Providing the services that people need is the mechanism for getting people from their automobiles onto transit service.

Jones also comments on the need for a new relationship with labor. The management-labor friction in the transit industry is legendary. There is much fault that one could find on either side. Jones suggests that a new contract between labor and management would be fundamental in allowing the public transportation industry to play the role he feels it needs to play for our urban areas to flourish.

Conclusion: Public Transportation

The view presented here is not particularly optimistic. Urban public transportation has fallen on hard times with some bright spots partially as the result of national policies here in the United States. From the federal public expenditure viewpoint, funds in the post-World War II era have been primarily allotted to highways and airports. The tax deductible status of the home mortgage and the American dream of a single family home has led to land-use patterns that are difficult for public transportation to serve. So public transportation has an uphill fight.

Some of the more innovative public transportation operators and administrators have recognized that a broadening of the definition of the very term "public transportation" and a focus on getting people out of SOVs is an appropriate strategy. Certainly our cities will be more viable if public transportation can accommodate to the current reality of finance and land use. But it will take many innovative and dedicated operators and managers to make that happen.

REFERENCES

1. Hoffman, A. "Toward a Positioning Strategy for Transit Services in Metropolitan San Juan," Thesis for Master of Science in Urban Studies and Planning, M.I.T., February 1996.

2. Gray, G. E. and L. A. Hoel, (eds.), *Public Transportation: Planning, Operations and Management*, 2nd Ed., Englewood Cliffs, NJ: Prentice-Hall, Inc., 1992.

3. McQueen, B. and J. McQueen, *Intelligent Transportation Systems Architectures*, Norwood, MA: Artech House, 1999.

4. Jones, D., "The Dynamics of Transit's Decline and Continuing Distress," *Urban Transit Policy: An Economic and Political History*. Englewood Cliffs, NJ: Prentice-Hall, 1985.

CHAPTER 29

Intercity Traveler Transportation: Air

....

Intercity Traveler Transportation

The coverage of intercity traveler transportation begins with air transportation, focusing mostly on the United States. From there we will consider rail transportation, again focusing primarily on the United States, but also looking at systems abroad.

....

Air Traveler Transportation: A Brief History

The Wright Brothers

The Wright Brothers in North Carolina in the early part of the twentieth century were the first to achieve heavier-than-air flight. Air transportation got a major boost (as do many innovations) through military use during World War I, during the 1914–1918 period. This demonstrated the long-term potential of the air mode. It became a mode that was viewed as actually useful for civil applications, rather than a curiosity (wing walkers and the like).

Airmail and the Kelly Act

In 1918, the United States began airmail service; in 1925, the Kelly Act permitted the Post Office Department to award airmail contracts to private companies. This is viewed by historians of transportation as a very significant event in the development of the United States air network. That same Kelly Act also required companies who carried the mail by air to carry passengers. As the federal government recognized the potential of airline passenger transportation, they used the revenues associated with moving the mail around as leverage to entice private companies into providing, not only airmail freight service, but passenger service as well.

In the 1930s, airmail contracts provided a substantial fraction of the infant air industry's revenue. And economists argue that the government paid a substantial premium above costs—that is, a subsidy to those companies—in the interests of developing an air industry. Also, in the 1930s the federal government assumed responsibility for the national airways system—the air traffic control system, as it is now called.

World War II

There was relatively slow growth of air transportation in the 1930s. Safety problems plagued the air industry. There was, of course, a worldwide economic depression that affected the United States, so there was not much discretionary funding for investments. Also, technology did not advance much during that period. But then, in the late 1930s and early 1940s, along came World War II, which provided the next major push for the development of airline transportation. There were extraordinary technological advances in aircraft design and manufacturing expertise. The airplane, as a major element in the conduct of World War II, advanced technologically.

There were also social factors that were coincident with World War II that affected the air industry. First, many people were exposed to flying who would not have been exposed under more normal (peace time) circumstances. People in the Armed Forces were transported by air. The air industry received some impetus through the fact that people became used to the idea that flying was not limited to the equivalent of astronauts of the 1940s, but rather was something that normal, everyday people could do.

Post-World War II

In the post-World War II period, the United States had a rapid economic expansion. There was a good deal of pent-up demand for automobiles, among other items, and there was income to pay for it. The technological advances that had been made during World War II greatly advanced the airline industry during this period of economic growth.

Commercial Jet Service

The airline industry grew at a rapid rate in that period. In the late 1950s, commercial jet service was introduced for the first time. Up to then propeller planes were used. Jets had increased range, and faster operating speeds; the comfort levels of people in jet airplanes were improved as well.

The Eastern Shuttle

In the early 1960s, there was an important marketing innovation—the development of Eastern Airlines' shuttle in the Northeast Corridor. (Eastern is now out of business, but the shuttle service remains.) The Eastern shuttle involved nonreservation hourly airline service between Boston, New York, and Washington.

The shuttle was a major change because it positioned the airlines as a mode for the people, rather than a mode for the elite. One could simply walk into the terminal and get on the airplane, and literally (in those days) buy one's ticket on the plane. No reservation was required. Eastern advertised and delivered on the promise that if there were more people for the 6 o'clock shuttle than could fit on the plane, then they would wheel out another plane, even if only 2 or 3 people required service. And the fact that it was a no-reservation, no-frills, on-demand service put it at the leading edge of airline marketing at that time.

Wide-Bodies

In the 1970s, the air industry continued to grow. We saw the advent of wide-bodied aircraft—747s, DC-10s, aircraft that had two aisles that allowed for as many as 11-across seating, depending upon the configuration, with capacities in the 400 range.

Airline Costs

Labor costs in the air industry are an important factor in overall costs. This includes flight crews, maintenance crews, ground crews, the people that sell you tickets at airports, and the like. On an industry-wide basis, on the order of 35 to 40% of the total costs in the air industry are labor-related.

Fuel is also a major component—on the order of 20%—of the total cost of running the United States airlines. Fuel costs tend to change quickly and substantially. During the oil shocks of 1973 and 1979, jet fuel costs went up dramatically and the impact on the airline industry was quite substantial, and negative.

Equipment costs are an important component of the cost picture, with large planes costing upward of $100 million. When one includes depreciation and interest charges (these aircraft are financed by the carriers), these are substantial costs. Therefore, high equipment productivity is critical. Keeping aircraft on the move and full and minimizing dead time is critical.

Regulation

The air industry was deregulated in 1978, allowing them more pricing freedom than had existed prior to that time. The airlines have substantial freedom to change their prices in response to competitive and market forces and operating situations. Entry barriers to new carriers were substantially relaxed.

Reasons for Air Industry Financial Problems

There are many reasons as to why the industry faces periodic financial problems. Competition is obviously one of the critical elements. There are those that would argue that the industry has more capacity than it needs for the demands it serves. There are a number of niche players like Southwest Airlines who are running a highly productive, relatively low-service-quality operation (from the point of view of the on-board services that a customer would receive)—providing very cheap fares and attracting passengers from the conventional carriers.

Earnings in the airline industry are very sensitive to the ratio of filled seats to total seats. Once a seat flies empty, the revenue from that seat is gone forever. And airlines, recognizing that fact, have gone through some destructive pricing battles. Highly discounted fares are available, particularly in circumstances where you are willing to purchase a nonrefundable fare a long time in advance. What this is reflecting is the airlines' recognition that filling seats, even with a low-priced customer is better than the seat flying empty. Some feel it depresses the market for the expensive seats.

The airline industry finds it difficult to quickly adjust its fleet size and hence its capacity. The time between ordering new aircraft from the manufacturer and delivery to the airline can be several years. This, combined with the difficulties in predicting the state of the air traveler market several years hence, suggests that aircraft ordered during periods when demand is strong may arrive when this is no longer the case, saddling the carrier with expensive unneeded capacity.

In international travel, many international carriers are subsidized by their national governments where, at least in principle, United States carriers are not. Therefore, the argument is that the competition that exists for the very lucrative international market is not on a level playing field. The United States carriers will argue that, because they are non-subsidized carriers, when Air France is subsidized by the government of France, competition is fundamentally unfair. What airline is permitted to provide service to which country is a rather political and contentious matter.

Now, we are seeing a number of carriers forming strategic alliances across international boundaries—US Airways and British Air; United and Lufthansa; Delta and Sabena; Northwest, KLM, and Asiana are examples.

Air Traveler Transportation and the 30 Key Points

Earlier, we introduced 30 Key Points in transportation. Thinking back to that list, which of them might have particular impact on the airline industry?

Stochasticity

The first point of particular importance to the airline industry is stochasticity (Key Point 26)—the fact that transportation systems are subject to a variety of probabilistic forces on both the supply and demand sides. Stochasticity in the airline industry—for example, with respect to weather—is a particularly important factor. The single largest cause of delays in the air transportation system is weather. If the weather is poor, the level-of-service will deteriorate.

Peaking in Demand

Another point that is particularly important in airlines is temporal peaking in demand (Key Point 18). The fundamental issue is how to choose capacity—how often don't we satisfy demand? The airline industry is subject to some really severe peaks—holiday travel, for example, around Thanksgiving and Christmas. Business travel takes place primarily during the week, with peaks early and late in the day, in much the same way as the journey to work takes place. Vacation travel is heavily skewed toward the weekend and there are seasonal peaks as a function of destination.

Selecting Capacity

Peakedness is of particular importance in the air mode. It relates to the fact that airport capacity—the ability of airports to land planes at some rate—is the major bottleneck. Landing and take-off slots at particular airports may be highly constrained. So, we have delays as a result of capacity limitations at airports.

   This is a serious issue because expanding airport capacity—at least by the conventional means of building new airports and new runways at existing airports—is very difficult in the United States. The pressures against building new airports in major metropolitan areas are quite severe. Everybody wants the airport, but nobody wants it near them because of noise considerations and other impacts such as automobile traffic. Assembling the land to develop an airport in a major metropolitan area is close to impossible. So, very few airports have been developed in this country in major metropolitan areas over the last several decades. The Denver International Airport is the only one that has been developed in the last 20 years.

Network Behavior

Another relevant point from the 30 Key Points is network behavior (Key Point 25). Networks behave in ways that are not easy to predict. The airline industry is particularly prone to these network effects; the air network operates in a highly interconnected way. The structure of the network is such that local delays can propagate easily to a broad geographical scale. The high degree of interconnectedness of the network is a defining characteristic of the air transportation system.

It is worth commenting in this context that the motivation for the new Denver International Airport was not simply better access to Denver. Rather, its role in the national air network is central. Weather in Denver is often poor, and the physical configuration of the old Stapleton facility limited operations in bad weather, thereby making Stapleton less than optimal as a hub.

Land-Side Issues

We continue with several land-side issues in air traveler transportation.

Airport Access

People are not interested in going from one airport to another airport, but rather are interested in going from some origin to some destination. So, we have a trip that involves, say, traveling from the center city out to the airport, flying to a second airport in the destination city, and having another link that takes you from the airport into that center city (see Figure 29.1).

So, the door-to-door transportation time has important components beyond the air link. We have a surface trip to the airport, and typically some access and slack time within the airport—you have to park your car and you have to walk through the terminal to your gate. So, you will leave a certain amount of slack time to be sure you don't miss your plane. You then have to "de-access" the airport: get off the airport property; rent a car or get on the airport bus, whatever it may be; and go to your final destination in the city.

FIGURE 29.1   *Airport location.*

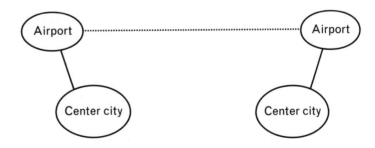

LOS Implications

The level-of-service that the customer observes is not determined only by the speed of the air link, which is the only service that is being directly provided by the air carrier. It is also a function of other characteristics of the trip, such as the distance to the airport from the center city, the congestion that one encounters going to or from the airport to the center city, and the like.

Some shuttle services, such as the Boston to New York shuttle, explicitly recognize the intermodal nature of the service they offer. Their service includes a transfer to Manhattan, for an additional fare at the customer's option, paid for when you pay for the air ticket.

These shuttles recognize what their service really is. Their service is, in fact, getting the passenger where he or she wants to go, and not simply getting them to the destination airport. Virgin Atlantic Airlines advertises picking you up by limousine to take you to the plane before you fly to Europe. So, the notion of an integrated intermodal service is becoming better recognized in the passenger transportation industry.

Proximity of Airport to Center City

The question of the proximity of the airport to the center city is important: total trip time is clearly going to be a function of whether this distance is 5 miles, 20 miles, or 50 miles. In the Boston metropolitan area, we have a very close-in airport, Logan Airport, which is a very short ride from Boston's downtown, under good traffic conditions.

Another example of a very close-in airport is National Airport (now Reagan National Airport) in Washington, D.C. National has the additional benefit of being accessible by rail transit on the Washington Metro. This is also true in Boston, although it is a three-transfer trip from MIT. Even so, at peak hours when highways are congested, it may be the fastest way.

The new Denver International Airport is a substantial distance from Denver. Internationally, Narita Airport, serving the Tokyo metropolitan area, is on the order of 50 miles from downtown Tokyo. JR East runs a good quality train—the Narita Express—that takes almost an hour to get from Narita to Tokyo Station (nonstop), in the center of Tokyo.

Rail Access to Airports

The notion of having access to airports by rail—which has long been available in European airports, for example, Frankfurt or Zurich—is now becoming important to major metropolitan areas in the United States [1]. Consider Hartsfield Airport in Atlanta; you are not even out of the terminal and you are in the subway—MARTA—taking you to the center of Atlanta for $1.50 (1999). You can access O'Hare by the CTA as well.

There has been discussion for years in New York City about developing a "train to the plane," where a train would go directly from the East side of Manhattan to La Guardia Airport. Having that kind of access through Queens at high speeds—certainly high speeds compared to the relatively congested highway links and the relatively slow subway service that exists now—is a possibility.

The people in Queens, which is the borough of New York City that is home to La Guardia, are concerned about the impact on their borough of these trains speeding through, with the people of Queens having no direct benefit from that high-speed service, because it would go, basically, from Manhattan to the airport and would not provide any service for the people in Queens itself. It is a classic, difficult political issue transportation systems face all the time. Who benefits and who is impacted negatively? You have to go through somewhere to get somewhere. Those who travel get the benefits. Those who are being "gone through" pay a price with no benefits.

Airport Terminal Design

Airport access is an example of a land-side issue. Another land-side issue is airport terminal design. How do you provide efficient terminals that are customer-friendly (e.g., minimizing walking)?

Baggage handling is a vital part of the land-side operation. Baggage handling, particularly in an environment in which there are many transfers of passengers among planes at airports—as is the case in "hub-and-spoke" operations, which will be discussed later—is very important.

There have been some studies done about the psychological importance of the airport as a gateway to an urban area: you form your first impression of the city from the airport. It is the first place you see when you get off the airplane. Many urban planners and urban developers claim that the airport is a psychologically important entry to the urban area, in much the same way that railroad terminals were, back in an earlier stage of the United States' development; one would look at Grand Central Station, Pennsylvania Station, Union Station in Chicago, as major gateways that really marked, in a physical and emotional way, the entry into a particular metropolitan area.

Airports as Commercial Centers

The idea of the airport itself as a commercial center is a very interesting one. In Japan, the East Japan Railway Company, which owns the stations, treats them as an important commercial opportunity. They operate the shopping areas within the stations. In many Japanese cities in the 0.5 to 1.5 million population range, the train station is the center of commercial life. People in Japan go to the rail station to do shopping in the same way you might go to a mall. It is the place where nice shops and restaurants are. JR East has used its stations as commercial profit centers. We are seeing this idea in airports as well, with upscale shops and restaurants.

Important Air Issues

Airport Capacity

As noted earlier, capacity limits in the airline industry are very closely tied to airport capacity. The abilities of airports to handle take-offs and landings, particularly at peak hours, one could argue, is the limiting factor in the growth of the airline industry, particularly given the fact that building new airport infrastructure is difficult.

Congestion Pricing at Airports

An idea for making better use of airport capacity is congestion pricing, which we have introduced in other modes. It can work, but there is also a tremendous amount of political resistance. Airlines pay landing fees to airports. One could charge the airlines different rates for landing at different times of day, and have the effect of "spreading the peak," as in other modes.

One of the targets of congestion pricing is general aviation, which are relatively small, privately-owned aircraft that land at major airports. If they land at peak hours, they cause major congestion and delays, and, often, don't pay a great deal of money to land, given that the landing fees are typically based on weight. So, why not charge general aviation higher fees at busy times to "force" them out of the peak?

The political impediments to implementing congestion pricing have proven to be very formidable. There was an experiment done here in the Boston area a few years ago, in which landing fees were charged that drove general aviation out of the peak by charging higher fees at peak times. Congestion was reduced substantially. Ultimately, though, the political and legal pressures proved to be irresistible, and congestion pricing was eliminated from Logan.

Hub-and-Spoke Airline Operations

Figure 29.2 shows a hub-and-spoke operation. Customers bound for three East Coast cities from three West Coast cities "hub" through Denver (see Figure 29.2).

Hub-and-Spoke as a Cost/LOS Trade-Off

This has become an important operating mode of the United States airlines; it represents the classic cost/level-of-service trade-off. The quality of service offered is less than a non-stop flight from the West Coast to the East Coast would have provided. But the cost to the airline of providing that hub-and-spoke service was certainly less than providing non-stop service. The idea, at least in principle, is that part of that cost-saving is passed on to the customer. In a highly competitive environment, as certainly exists in the airline industry, that is probably true. Direct service is expensive; hub-and-spoke service is a less expensive, lower quality-of-service at a lower cost to the airline and, in principle, the passenger.

FIGURE 29.2
Hub-and-spoke air network.

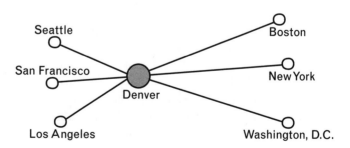

There are other implications of operating a hub-and-spoke operation. First, we mentioned airport capacity as a capacity choke point in the United States national system. Hub-and-spoke operations tend to exacerbate that kind of capacity problem. There are a lot more airplanes flying in and out of airports, and they are scheduled to minimize intermediate ground time for passengers. Therefore, take-offs and landings tend to be bunched—high peak loads result.

Hub-and-Spoke Operations and System Stability

We introduced the notion of system stability in discussing freight railroads. Let's consider this in the air context. Suppose we have tightly bunched take-offs and landings. If everything is running smoothly, people receive a high level-of-service, because the interval between landing on the inbound and taking off on the outbound is relatively short. But, if bad weather occurs, or if a plane is delayed for equipment reasons, or any number of other possibilities that may occur, the system can become unstable very quickly in this kind of tightly-scheduled operation.

You have the same issues you consider in railroad freight operations. Do you hold for traffic, for example? What are the implications of holding for traffic? Do you hold the Denver-to-Boston flight for me coming in from San Francisco, where, for whatever reason, there were departure delays out of San Francisco? The Seattle and Los Angeles flights are already there, and the people are sitting there on the Boston flight saying, "Why don't we take off?" Does the airline say, "We are waiting for the San Francisco flight?"

If you did, you provide me, coming from San Francisco, with a better level-of-service, because I don't have to wait for the next flight, which may be tomorrow, if it is late in the day. On the other hand, you are providing a poorer level-of-service to the people who are already in the plane from Seattle and Los Angeles, and you may, if you delay that flight, be causing a variety of problems in Boston, where they may be depending on the timely arrival of that aircraft for dispatching to another destination.

So, a hub-and-spoke operation is finely tuned. When it runs like clockwork it runs very cost-effectively, and provides a reasonable trade-off between costs and level-of-service. But the system can go out of kilter relatively quickly. The airline industry has high stochasticity, so that is not an uncommon occurrence.

Network Control: Ground Holds

"Ground holds" are an important part of modern air network control. In earlier times, you would take off from, say, Boston, and you arrive in the vicinity of New York, and you would circle, waiting for air traffic to clear, so you could land.

Current operations are rather different. Now we wait on the ground at the origination airport until a landing slot is available. Why? First, it is dangerous to have all these planes in a limited air space circling an airport waiting for their turn to land. It is a lot safer to have them sitting on the ground somewhere, rather than circling. Second, it is more economical to have them sitting on the ground, because they are not burning fuel as they circle. The time in the air is minimized.

Now, the system is not flawless by any means, although it certainly is a substantial improvement. In particular, one of the reasons for ground holds is weather in the city to which you are going. Weather is not always easy to predict; whether the weather will clear in one hour or two is beyond the state of the art in modern day meteorology. The implication of this is you may waste capacity. You are sitting on the ground in the origination city, and it clears earlier than expected at your destination. The person in the control tower at the destination says, "I'm all clear." He looks up in the sky (figuratively), and there aren't any airplanes there to land, because nobody was taking off to land there for the last several hours, because bad weather was expected to persist.

This system of ground holds and more generally of controlling the air network is often called the flow control system. It is an attempt to provide optimum air network operation, balancing safety, efficiency, profitability, and level-of-service, in a highly probabilistic environment with competing airlines and various (and sometimes competing) airports.

Safety

Safety is a key level-of-service variable in the airline industry. Although substantially safer than automobile transportation, the fact that the accidents that do occur are big and eye-catching makes travelers very sensitive to safety concerns. The Federal Aviation Administration (FAA) has responsibility for regulating airline safety; they also have a charter to promote the airline industry. Some feel there is an inherent conflict in these two roles.

Aircraft Technology

Aircraft Size

Aircraft size is an important parameter in the airline industry. The development of larger planes has been a continuing trend in the passenger business. Earlier we noted the development of wide-bodied aircraft, with capacities on the order of 400, in the early 1970s. Even larger aircraft—and the term that is used now is VLA, or "very large aircraft"—are in development. The idea of technologies that would allow economic transportation of, say, 800 people in a single aircraft is being explored.

The theory is that there are economies of scale in this kind of operation, just as there are in oil tankers, as discussed previously. In very large aircraft, you can transport people at lower cost per seat-mile. VLAs have many implications. From the point of view of airport congestion, what would be your guess as to the implications of VLAs?

It would relieve airport congestion.

Why is that?

Because it would have more people coming in per airplane, lowering the number of operations.

That's right. Rather than having two airplanes of size 400, you have one of size 800. But what else?

You might have some severe land-side problems if airports are geared to 400 rather than 800 people on a flight.

This might require some re-engineering of the land-side. With 800 people coming in at 2:00 p.m. from London, 800 from Paris, and 800 from Brussels, and everybody wants to clear customs at Logan, it could be a difficult situation.

What about level-of-service to the traveling public? How do you think very large aircraft would impact that?

If you have incoming passengers that still have to pass through one gate, it is going to take you much longer to get off the plane.

Correct.

There would probably be fewer flights.

Exactly. I would suspect that the airlines will fly fewer flights with 800 capacity. Rather than five times a day service to Chicago, maybe they'll only fly three times. So, as a customer, it may not look quite so good to me. I have less flexibility. The airlines costs may be lower and, in principle, that will be reflected in my price. But, indeed, there are some fundamental level-of-service/cost trade-offs there.

Short Take-Off and Landing Aircraft

Some other innovations in the airline industry: we noted jets coming in the 1950s, wide bodies in the 1970s, and maybe very large aircraft in the future. Another technological innovation is so-called "short take-off and landing aircraft," or STOL.

The idea here is that two of the problems with airports are distance from center cities and capacity problems. These both result from the long flight path needed to either land or take off. You land gradually, over a period of time; it takes a long distance to stop once on the runway. This implies huge amounts of land for airports and a big noise footprint for airplanes. That is why airports are often built far outside the city they serve.

STOL would configure an aircraft so it could position its engine vertically. It could act like a helicopter and take off vertically. Then, once aloft, it would position the engine horizontally to fly normally. Then at the destination city, it would land vertically. The implications for airport location and capacity are potentially substantial.

Whether this will be achievable within cost and safety parameters is still open to question. There has been a lot invested in research and development in the area, and the military is, of course, very interested in it. Whether it will have civil application is not yet clear.

Hypersonic Flight

Hypersonic flight is potentially a very important idea. The Concorde gets you to Paris and London from New York on the order of three hours, but it has some problems. First, it costs a lot. It is a very high-cost, high-priced service, with a fairly limited market. It also causes sonic booms over land. But, certainly, going from Paris to Boston or New York in three hours has value.

The Space Plane

For the next generation, some of the visionaries in the airline industry are talking about the space plane, where flights would actually go up into suborbital space—up at 100,000 or 150,000 feet—and from New York to Tokyo in 2 or 3 hours.

Engine and Materials Technologies

There are engine and materials technologies vital to aircraft development—we won't deal with these here, but recognize that air transportation is a technology-intensive industry. It is an industry that, over the years, has received very positive boosts, from R&D done in support of the military, for World War I and World War II and in the Cold War.

Airplanes as a United States Export Industry

Connected to this large R&D investment, note that airplanes are a huge export market for the United States, with companies like Boeing. Airplanes as a key United States export (as well as a domestic) industry are very important.

Yield Management in Air Transportation

There are important commercial advantages behind trying to fill seats optimally on aircraft. How does one build algorithms to sell seats at various prices over time for a particular flight, to try to optimize revenue?

In yield management, the optimal use of capacity to maximize revenues, the airline industry is the leader relative to any other transportation modes. Here is how they think about what size aircraft to use on a particular route. The larger the aircraft, the higher the operating costs; however, the larger the aircraft, the smaller the spill-costs—the lost revenue because the plane was full. (Of course, we have to make an assumption of how much revenue we recover when passengers simply take the next flight.) The trade-off is illustrated in Figure 29.3.

FIGURE 29.3
Yield management.

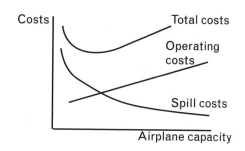

Air Transportation as an Example of Subsidies

We now use air as an example of a concept that is universal in transportation: the concept of subsidies. The question is how the cost to provide particular transportation services relates to the price to the customer for that service. In the air mode, there are different kinds of subsidies. Let us start by discussing intra-firm subsidies—subsidies that exist within an individual airline that is providing services to different kinds of customers and may, in fact, be subsidizing some of them through revenues received from others. Airlines receive revenues from some customers that are, in fact, subsidizing other customers.

There are several different kinds of subsidies we can discuss. The first is subsidies between long-distance and short-distance passengers. Cost functions look different for long-distance and short-distance passengers. On a per-mile basis, long-distance travel is cheaper to provide than short-distance travel. Long-distance passengers may pay a rate above cost, in effect, subsidizing the rates that are being charged to short-distance passengers.

A second subsidy is that between business and non-business travelers. Business travelers require flexibility to make plans on very short notice and change their plans very quickly. The airline industry charges them a premium for this service. Vacation travelers typically have more flexibility and make travel arrangements well in advance. The airline often provides cheap fares for these trips. There is a subsidy being provided by the business travelers to vacation travelers.

Another example is subsidies among various origin-destination pairs. One can have origin-destination pairs of comparable length and comparable cost to the airlines, where for competitive reasons, the fares are different. Customers on the non-competitive routes subsidize those on competitive routes.

Frequent Flyer Programs

What about the economics of frequent flyer deals?

The idea is to try and build some brand loyalty. You have a lot of miles on a particular carrier, and given it is a reasonable option in a particular case, you take that carrier.

It is another kind of subsidy. People who have frequent flyer miles get free travel. There is a subsidy between these frequent travelers and those who rarely travel, we could argue.

How about class of service—first class, business class, economy?

First class passengers are subsidizing the coach passengers. The actual costs of providing first class can't possibly be as much as the premium that people pay. However, some reasonable fraction of first-class travelers are doing it without paying cash. They are doing it by using frequent flyer benefits from previous trips to upgrade to first class.

Subsidies Resumed

We have talked about intra-firm subsidies—one customer subsidizing another. One can also talk about the broader question of subsidies among airlines, governments, and passengers, as shown in Figure 29.4.

Commercial airlines have passengers that pay fares for various kinds of services. On the right of Figure 29.4, there is the air system—the airports and the air traffic control system—basically the infrastructure of the air business.

The airlines pay toward the infrastructure through landing fees for using the airport. The airlines are paying money for infrastructure through various taxes, like taxes on jet fuel or even corporate

FIGURE 29.4
Flows of funds in air transportation.

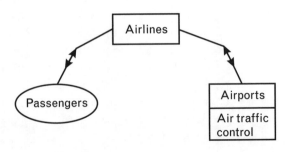

income tax. However, these funds are not enough money to support the system.

So we need another contributor, which is society-at-large (see Figure 29.5).

That is, the general tax revenues in the United States contribute to the maintenance of airport and air traffic control infrastructure. One can argue that the system in its entirety is subsidized by non-users, since money comes from the general tax funds of the United States, paid into by citizens and businesses that support the airports and air traffic control system of the United States. Although the airlines do pay for using airports, and passengers do pay for using airlines, and passengers do pay through taxes for the development of the airport and air traffic control infrastructure, and airlines through fuel taxes do pay to support airports and air traffic control, there is a general subsidy by society of the air system.

Does Society-at-Large Benefit Enough to Warrant the Subsidy?

The question is, does society-at-large benefit enough from its support of airlines and air passengers, and of airport and air traffic control infrastructure to warrant that subsidy? Now, through the political process, we have made a judgment that, in fact, we do benefit enough.

Now, air transporation is simply an example of subsidies. All modes of transportation in the United States are subsidized. All modes of transportation do not pay for themselves out of the "farebox" in a direct sense. Automobile users, public transportation users, intercity rail passengers—all are subsidized. There are invariably monies flowing from other sources that help provide for society's mobility. And there is nothing inherently bad about that. It can be argued that transportation is too important to fundamental societal needs to be subject to the vagaries of the marketplace.

FIGURE 29.5
Subsidies in air
transportation.

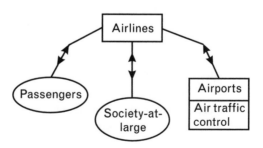

However, these subsidies lead to individual decisions that may be suboptimal from a societal viewpoint. When a commuter decides to drive to work, she doesn't bear the full cost. This may lead to suboptimal use of the overall resources in society. However, through the political process, society has basically made a judgment that the subsidies are appropriate and that the way in which we are allocating resources is appropriate, given the benefits we all derive from mobility, even those of us who may not be actually traveling.

The MBTA in Boston receives about 33 cents of the operating cost dollar from the farebox. (Note that this does not include any capital costs.) Society has made a political judgment to do that; it is worthwhile for the health of the city of Boston and Massachusetts and New England for people who ride the MBTA to be subsidized because it generates jobs and mobility and benefits for everybody.

Some of this has a social equity basis. People do have to get to work. If we are not providing transit service, how do people who are not well off financially and can't afford an automobile, participate in our economic system? In our car-oriented system, that is a very difficult thing for the poorer people in our society. That is one of the reasons that we, within the political process, subsidize public transportation systems.

There is another perspective on this, going back to the air example. By subsidizing the air system from the public treasury, one is generating the need for further airplanes sold by Boeing, for example, to the airlines. There are some benefits to be derived from that as well as for society-at-large. So these issues—jobs and industry viability—can be in the equation. It depends on how broadly you draw the boundaries around the system.

Consider again the elevator system introduced earlier. In that case the elevator system was totally subsidized by the building-at-large. Nobody paid to use the elevator directly; companies whose people used the elevator more or traveled longer vertical distances than other companies didn't pay any more for that use. The costs of the elevators were captured in the rent structure of the building. The owner computes rents as a function of square footage, but there are subsidies between the renters.

Perhaps one of the companies on the fourth floor is running a messenger service; people are constantly going in and out and using the elevator a great deal. Then there is another organization in which people simply come to work in the morning and go home at night. There is some subsidy going on there because we don't collect a fare every time somebody uses the elevator and use those funds to lower rent. So, even in that simple system, there are subsidies; we don't even really think about those subsidies because we have been conditioned to believe that that elevator service in a building is free. Of course, it is not—and so there are subsidies in that system.

REFERENCE

1. Coogan, M. A., "Comparing Airport Ground Access: A Transatlantic Look at an Intermodal Issue," *TR News* No. 174, November–December 1995.

CHAPTER 30

Intercity Traveler Transportation: Rail

Rail Traveler Transportation

We now consider rail traveler transportation. The comfort levels of properly designed and outfitted trains can be high—in many cases much more comfortable than riding in airplanes. The mobility while traveling, sitting down at a table and having dinner, sleeping in a legitimate bed, are all benefits. The fundamental problem is that you do not get where you want to go very fast. And, in today's society, getting there in a relatively short amount of time so you minimize the nights away from home drives intercity transportation, at least for the business traveler. So, despite what may be better creature comforts and a more pleasant trip in riding on trains, if your interest is in getting home and back as quickly as you can, it is difficult for trains to compete over long distances.

Rail Terminal Locations: An Advantage

An advantage of rail that exists to this day is, unlike airports, railroad terminals (many built in the nineteenth century) are in the center city (see Figure 30.1). The inherent advantage that rail has is that the access time to the center city from the station is much less than it is for air, given that airports are far removed from downtown. The trains were here first. The terminals are downtown. This is becoming less of an advantage, as origins and destinations have dispersed with the trends toward suburbanization. A much smaller fraction of trips are center city to center city as compared to forty or fifty years ago; therefore, the

FIGURE 30.1 *Rail station versus airport location.*

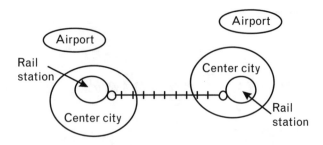

advantage of having these terminals in the city center is less than it was, although it is still a competitive edge.

Trains: A Surface Mode

A second major advantage of trains over air is they run on the surface. They represent a totally different set of energy and control challenges than airplanes do. You do not have to lift tons of aluminum off the ground, as you do with airplanes, which is very costly, from a structural design as well as from a fuel point-of-view. Further, from a control perspective, with trains you have vehicles running on tracks; this is a very tightly controlled situation, as opposed to vehicles flying in the air with no physical guideway.

With trains, we also have the ability of being able to link vehicles together to provide high capacity. If we find that the 8:00 a.m. train to New York is full every day, it is a relatively minor operating change to add a car to that train. Doing something comparable in the air mode typically involves flying another plane, which may be difficult if airport capacity is constrained.

Another advantage of trains that stems from the fundamentals of the mode is that weather tends to be much less of an issue. Certainly, major snowstorms can put trains into a difficult operating situation, but the sensitivity of a surface mode to weather will generally be less than that of an air mode.

The Problem: Speed

There are, of course, at the same time many disadvantages of the train system. The primary one is the obvious one—speed. They do not go as fast as airplanes.

The United States: A Big Country

Consider a country, a stylized United States looking like a big rectangle three thousand miles east to west, and 1,500 miles north and south (see Figure 30.2).

It is a big country and even with velocities of 200 mphr (which they are not), it takes a long time to get from place to place. The idea of a rail system—coast to coast and border to border—is not viable in the United States. Rather, rail as a regional solution, as in the Northeast Corridor, may be viable.

Noise Impact

Noise can be an issue in rail, but in a different way than air. Certainly air travel causes major noise problems in and around airports. But that is a relatively local impact. In many cities, the airports are quite a distance from the center city, and while people are impacted by noise, it is a relatively small number of people who are impacted relative to the population of the urban area at large. (This is not true in Boston, New York, or Washington with their close-to-center-city airports).

With trains, the noise profile is very different. It makes noise along its entire right-of-way. (So do airplanes, of course, but they're 30,000 feet in the air and you do not hear very much.) So we have a train going from Boston to New York, impacting people along the entire corridor. Abutters are very concerned about increasing the speed or the number of trains per day. So we have the noise problem with trains being a surface mode. "You have to go through somewhere to get somewhere." With air, that is less important.

The fact that not everybody is benefiting from the mode makes political concerns with noise and land-use along the right-of-way that much more difficult. For example, people in Connecticut are concerned about high-speed trains going from Boston to New York when,

FIGURE 30.2
"Stylized" United
States.

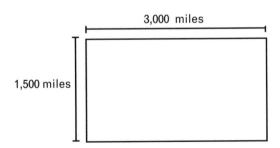

in fact, there are few high-speed rail stops in Connecticut because stops raise the total trip time. So the problem is that while impacts may occur along the corridor, the benefits do not accrue to those suffering the impacts.

United States National Transportation Policy and Passenger Rail

Now, the United States' national policy—since World War II—has not been supportive of rail systems for travelers. Public policy on the national level has focused on highway systems and air transportation systems and very little on intercity rail traveler service.

Reasons for the public policy emphasis on highway and air are the following. First, investments in highways in the post-World War II period were viewed as an important economic development strategy. It was a way of pumping money into the infrastructure to improve mobility around the country. It was a way of expediting and encouraging the conversion of United States industry from a war footing back to the production of automobiles. There was an enormous amount of pent-up demand when people could not find consumer items to spend the monies they were earning during World War II. Encouraging them to buy cars was a great way to jump-start the economy. So investment in the highway mode could be justified as an economic development strategy. Also, the highway construction programs themselves were helpful in generating economic activities and growth.

Second, the investment in air could be justified on a simple basis: as noted above, the United States is a big country. This country is 1,500 miles by 3,000 miles; if one is going to think in terms of a national transportation system that links together the whole country, air is really the only viable mode. So there were investments in air infrastructure—airports and air traffic control.

Third, if one goes back to the post-World War II era, environmental concerns were not a major factor in transportation decision-making. We have discussed the impact of automobiles on the environment. We think these concerns were always there, but if one goes back to the pre-1970 era, except for a few pioneers, the Rachel Carsons of the world, nobody was particularly worried about the environment. There were occasional articles about air quality, but consciousness was much lower than exists currently, where environmental impact statements are required for any major projects, and where clean air legislation affects transportation decision-making in fundamental ways.

If environmental concerns had been as important back then as they are now, one could have imagined a different United States investment policy. Investing in rail, which certainly is environmentally less intrusive than investing in highway infrastructure, might have been higher on the public agenda.

A fourth reason is that the railroads weren't all that interested in traveler transport. They saw freight as their business. They saw travelers as a money loser, and vocal in the bargain—people complaining about trains being late. It got in the way of their freight business, which was their core business. Fares were suppressed by government regulation. The railroads were not unhappy to see this whole line of business deteriorate. Recall that the United States railroad industry is composed of private profit-making corporations. The industry is not the instrument of public policy that railroads are in, say, France. The United States railroad companies were happy to divest themselves of intercity travelers, and to this day do not view it as an important part of their business. They are freight companies.

United States Intercity Transportation Investment by the Public Sector

Now, we have discussed air and rail as intercity traveler modes. Also, we have noted the dominant importance of private automobiles. These modes are different in the services provided, and from a financial and institutional perspective. In principle, we would like to think of an intercity traveler transportation system, composed of cooperating modes, but in practice, it is difficult in the federal and state transportation sectors to make institutional and investment trade-offs among modes.

So now, with environmental concerns much higher on the public's agenda than they were in the past, and with real congestion on the air system and a lot of congestion on the highway network, one might reasonably decide that what would really make sense is an investment in rail transportation—not as a national system, but rather as a set of regional systems for certain highly populated corridors. Rail investment would lower the environmental impact of transportation. It would have the effect of lowering congestion levels in air and highway. Why do we not do that? Why do we not invest in a high-speed rail (HSR) corridor between, say, San Francisco and Los Angeles, where one could make a pretty persuasive environmental case?

Strong Modal Orientation

The issue is partially institutional; the Federal Department of Transportation has very strong modal administrations. The Federal Highway Administration is particularly strong, but so is the Federal Aviation Administration. The institutional machinery to consider that modal trade-off discussion is not present. Parallel issues exist in Congress; different (and powerful) committees deal with different modes.

In the New England region the idea of having a second airport to supplement and relieve Logan has been on the table (off and on, actually) for a number of years. According to some projections, Logan will become highly congested in a few years as air traffic grows. A second major airport (as well as Logan expansion) has been considered, but siting such a facility is very difficult.

At the same time, rail advocates suggest high-speed rail between New York and Boston. Some substantial fraction of air operations in and out of Logan are associated with the Northeast Corridor—with flights to New York, Washington, Philadelphia, and so forth. By providing high-speed rail, we could provide an alternative transportation mode that would push the need for another Boston area airport many years into the future—certainly further out into the future than it is now. Whether the three-hour Boston-New York trip (reduced from four-plus hours), now scheduled to begin in 2000, will attract many more riders remains to be seen.

Comparable arguments could be made for investments in rail rather than in highway infrastructure. But there are not any real institutional mechanisms for considering this as an intermodal system issue. The public funds are not readily transferable among highway, air and high-speed rail.

Rail Passenger Data: A Historical Perspective

We now consider some rail traveler data to give you some historical perspective (see Table 30.1). Back in 1930, the air industry was in its infancy, and the highway system was at an early stage of development. Rail was a venerable mode. Consider the market shares for travelers

TABLE 30.1   United States Passenger Market Share (Commercial Carriers (%)) (*Source:* Wilner, F. N., The Amtrak Story, Omaha, NE: Simmons-Boardman Books, Inc., August 1994.)[1]

YEAR	RAIL	BUS	AIR
1930*	74.6	18.1	0.2
1940	68.5	28.2	3.3
1945	74.7	21.9	3.4
1950	50.4	35.2	14.4
1955	39.9	30.5	29.6
1960	29.8	26.5	43.7
1965	18.5	25.0	56.5
1970	7.5	17.3	75.2
1975	3.9	14.7	79.4
1980	4.5	11.3	84.2
1985	3.6	7.6	88.8
1990	3.5	6.0	90.5
1991	3.6	6.3	90.1

*Percentages for the year 1930 do not total 100 percent because waterway travel still attracted a substantial percentage of commercial travelers.

among the three commercial modes: rail, bus, and air. In 1930, the rail industry had about three-quarters of the market; bus was about eighteen percent; and air was about two-tenths of a percent. In 1956 the Interstate highway program began, but we had already seen substantial deterioration of the rail market share by then.

Today, about 90% of nonauto travelers go by air, so air has become the dominant mode over the sixty-year period shown, while rail has become marginal. (This is the share among those who are traveling by commercial carriers. We emphasize that we still have a majority of people choosing to make intercity trips by automobile).

Why did passenger trains become less important? We have already touched on several of the reasons. Air transportation became a strong competitor, as did automobiles with the development of a strong highway system. Both the air and highway modes were federal favorites in

1   This book is an excellent history of the Amtrak system.

the post–World War II period. A further reason was the strong labor union protection in the railroad industry prevented the railroads from rationalizing their costs, making them an expensive provider of transportation services. Also, as mentioned earlier, the rail industry felt their future was in freight transportation and were not unhappy to see rail traveler transportation deteriorate.

Amtrak

In the early 1970s, Amtrak was formed by the United States federal government as a quasi-public organization to provide traveler rail service on an intercity basis around the United States. Amtrak has had a mixed record over its history. Table 30.2 shows data on route-miles, travelers carried, and traveler-miles from 1972 to 1993.

Amtrak has run an operating deficit in every year of its operation. Amtrak owns and maintains its own rolling stock and stations. The rights-of-way, however, are provided by the private freight railroads. Under the legislation that brought Amtrak into being, the freight railroads must provide Amtrak access for traveler service. Amtrak pays the railroads for use of their right-of-way.

The responsibilities for dispatching trains are retained by the freight railroads. Given their perspective on where their real business interests lie, you can imagine what happens when a choice is to be made between a passenger and a freight train. Amtrak does pay incentives to the railroads for on-time service and recently has been more assertive in insisting upon reasonable access to the right-of-way to provide high-quality traveler service. We will return to Amtrak later in this chapter.

Traveler rail service in the United States is not currently a high-speed business. The fastest trains in the United States are on the Washington/New York portion of the Northeast Corridor, where speeds of about 125 mph are attained. Compared with rail systems in other countries, this is slow indeed. High-speed rail is currently under study in several densely populated corridors around the country.

The Federal Railroad Administration is currently conducting research on high-speed rail, including some on diesel locomotives that can sustain 150 mph speeds (recognizing much of the United States rail system is not electrified) on advanced train control systems and grade-crossing protection.

TABLE 30.2   Amtrak Profile (*Source:* Wilner, F., The Amtrak Story, Omaha, NE: Simmons-Boardman Books, Inc., August 1994.)

YEAR	ROUTE MILES (000)	STATIONS SERVED	INTERCITY PASSENGERS (MILLIONS)	PASSENGER-MILES (BILLIONS)
1972	23	440	16.6	3.0
1973	22	451	16.9	3.8
1974	24	473	18.2	4.3
1975	26	484	17.4	3.9
1976	26	495	18.2	4.2
1977	26	524	19.2	4.3
1978	26	543	18.9	4.0
1979	27	573	21.4	4.9
1980	24	525	21.2	4.6
1981	24	525	20.6	4.8
1982	23	506	19.0	4.2
1983	24	497	19.0	4.2
1984	24	510	19.9	4.6
1985	24	508	20.8	4.8
1986	24	491	20.3	5.0
1987	24	487	20.4	5.2
1988	24	498	21.5	5.7
1989	24	504	21.4	5.9
1990	24	516	22.2	6.1
1991	25	523	22.0	6.3
1992	25	524	21.3	6.1
1993	25	535	22.1	6.2

International Systems

Some international systems provide high-speed, high-quality, nationally-scaled service. The TGV in France routinely provides service approaching 200 mph. Shinkansen operations provide service in the range of 170 mph throughout Japan and have operated since 1964. High-speed rail technologies in Germany (the ICE train), Sweden (tilt trains), and Italy are also deployed. These countries have all made strong commitments to high-speed rail as a viable alternative for domestic air or highway for intercity travel. None of these countries has a highway system or an air system comparable to that of the U.S., and all of these countries are substantially smaller from a geographical standpoint than the United States. Therefore, the high-speed trains in these nations can span the country in a much shorter period of time than is possible in the far-flung United States. (There is also much current discussion about a trans-European high-speed rail network.)

France, Japan, and Germany are all aggressively marketing their high-speed rail technology to other nations. South Korea has made a commitment to TGV technology for their Seoul to Pusan route. Taiwan has been discussing a high-speed rail corridor from Taipei to Kao-hsiung for some time.

TGV has had some interesting prospects in the United States, including one, the so-called Texas Triangle Project, that was to be built by Morrison-Knudsen connecting the three major cities in Texas—Houston, Dallas, and San Antonio—but this project failed for various reasons, including financial problems brought on by the fact that it was largely a private-sector initiative. The successful high-speed rail projects in other countries are all public-sector dominated. The ability of the private sector to run high-speed rail on a for-profit basis anywhere in the world is open to substantial question. However, one could, as public policy, create a subsidy system that would make high-speed rail viable. The United States has not seen fit to do this—for better or worse. The state of Florida had a major plan for a Miami-Orlando-Tampa HSR system, as a joint public/private venture of Florida and the private sector, but in 1999 the state withdrew.

Technology for High-Speed Rail

The technologies for high-speed rail used in Japan, France, and Germany all require a dedicated right-of-way (no other passenger or freight rail service). Track structures are typically of continuous welded rail and concrete ties. Due to design speeds, there are horizontal and vertical curve constraints that are much more stringent than for conventional trains.

For power, electrification is standard. Electrification is hardly standard on the United States rail system. Rolling stock for high-speed rail uses low-weight equipment, since energy costs are proportional to the weight of the car and to the cube of speed. Also, aerodynamic effects are very important at high speeds.

Noise becomes more of an issue with high-speed trains. For example, the noise of the pantograph on the top of the cars picking up electric power from power lines is quite substantial at high speeds. This has been an issue in Japan where much research is being done on pantograph noise, and is an issue in the United States Northeast Corridor, where noise along the right-of-way is a concern.

Some rolling stock technology of interest, particularly on the Northeast Corridor, are so-called "tilt trains." Tilt trains allow for high speeds retaining traveler comfort on curvy rights-of-way—of the sort that exists on the Northeast Corridor. There have been various experiments with tilt trains, including Asea Brown Bovari (ABB) and TALGO technology, on the Northeast Corridor. Bombardier has been selected to provide high-speed train-sets to Amtrak and, as noted earlier, three-hour service from Boston to New York is scheduled to begin in 2000.

Another technological area is signaling, communications and train control, which become more critical as train speeds increase. Safety at grade crossings is a major concern with high-speed traveler trains. The concern here is not only the highway vehicle that might be in an accident, but the large number of train travelers that would be in danger in an accident situation. In France, the grade-crossing problem for high-speed rail was solved very directly. There are no grade crossings. Whether this can be achieved in the United States, if high-speed rail were to come to portions of this country, is less clear.

The Cost of Speed

A key concept: speed costs money. Maintenance costs go up dramatically as operating speeds increase. This is shown in Figure 30.3. The maintenance costs per mile are about four times as high for a track maintained to 120 mph standards as they are for a track maintained for

FIGURE 30.3
Maintenance cost versus speed (Source: In Pursuit of Speed: New Options for Intercity Passenger Transport, Special Report 233, Transportation Research Board, National Research Council, Washington, D.C., 1991).

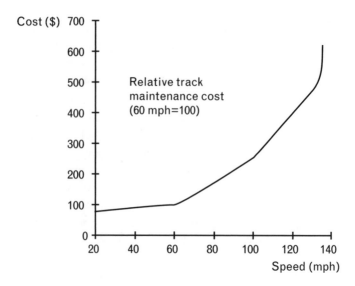

60 mph standards, and continue to go up dramatically as train speeds increase.

Incremental High-Speed Rail

Recognizing the difficulty with supporting dedicated high-speed rail service in the United States, an approach that is being considered now in this country is called "incremental" high-speed rail. Daniel Roth did a careful study of incremental high-speed rail [1]. The basic concept is, rather than having dedicated right-of-way for high-speed rail services, there would be an upgraded but shared right-of-way with both traveler and freight service that would be operated at higher speeds than currently available—perhaps in the 130–150 mph range (for the passenger trains). The costs in upgrading track to permit these higher speeds would be substantially less than building a dedicated line.

Incremental high-speed rail would provide some traveler rail services in the United States that are substantially better than we have today, and perhaps at reasonable cost. However, the freight railroads, in general, are not enthusiastic about the idea of sharing the rights-of-way they own with more passenger trains. The development and deployment of better train control systems to enhance safety in this mixed use would be necessary.

Mag-Lev

The current state-of-the-art with Japanese or French-style high-speed rail, that operates on a dedicated right-of-way in the 150–200 mph range, is steel-wheel on steel-rail technology—by using better suspension methods, better control systems, and so forth, these speeds can be improved incrementally. However, if one wants to think about non-incremental change, there is the technological concept called "mag-lev," which stands for magnetically-levitated systems. The idea here is using very powerful magnets, based on super-conducting technology. A vehicle with magnets runs along a magnetized guideway with tolerances between the vehicle and guideway on the order of fractions of inches. In effect, the vehicle "flies" above the right-of-way with no physical contact (see Figure 30.4). Richard Thornton is one of the early pioneers and major proponents of mag-lev technology as an important concept for surface transportation [2].

Speeds of potentially 300–400 mph plus, well beyond what one could achieve with high-speed rail (steel-wheel on steel-rail), can be attained. Of course we need dedicated rights-of-way, and indeed, very expensive rights-of-way; we need super-conducting magnets and need to maintain the infrastructure to small tolerances to ensure its proper operation, given the small air-gap this technology depends on.

FIGURE 30.4
EMS attractive mag-lev system and EDS repulsive mag-lev system (Source: Phelan, R. S., "Con- struction and Maintenance Concerns for High Speed Maglev Transportation Systems," Thesis for Master of Science in Civil Engineering, MIT, June 1990).

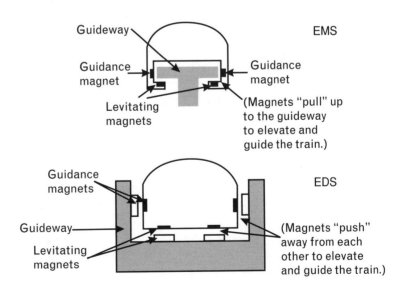

This is a possible next step in surface transportation. It is a high-tech solution. It is a complex technological system including difficult issues in guideway design. There are currently no operational systems in commercial service. In Japan, mag-lev service between Tokyo and Osaka is being considered; this would reduce the travel time from 2.5 hours (by Shinkansen) to under an hour.

There is much difference of opinion among transportation experts about the potential usefulness of mag-lev. Some feel that it is truly the wave of the future, and that in 50 years this will be the international standard, and the United States is missing a major bet in not doing the R&D necessary to project itself into the international marketplace.

On the other hand, there are equally well-informed people who believe it will never happen. They think that it will never be economically feasible; they believe that the high capital costs make it a very unlikely prospect.

Mag-Lev Service Concepts

Now, there are many ideas about mag-lev deployment in the United States. Some envision a system with a national network, not unlike the Interstate or the air transportation system. With velocities of 400 mph or higher being feasible, the idea of coast-to-coast trips may be reasonable. You will need national political support to make this happen. You might have to put in some relatively low-density lines through states like Wyoming and Montana to get the national backing to produce the federal capital necessary to develop a system of this sort. That is one point of view on mag-lev.

Another concept is mag-lev as a feeder system for the airlines, and as a reliever system for relatively short-haul airline travel. The idea here is that mag-lev can be quite competitive in 500- or 600-mile trips. Given access time to the airport, one could argue that surface modes could be quite competitive for relatively short-haul service, and that the big market for mag-lev (or high-speed rail) would be in the 400- to 600-mile trip length.

In this concept, mag-lev is not a national system, but rather a system that would interface directly with air service through access to airports, and would compete only in the short-haul market. Rather than flying, for example, from Pittsburgh to Cleveland, and then Cleveland to the West Coast—if you were hubbing out of Cleveland—rather, you would take a mag-lev vehicle, at 300 mph, from downtown Pittsburgh to the Cleveland airport and access the air system to take an airplane to Houston or San Francisco. Also, mag-lev as a short-haul system simply providing airport access is being studied.

Incremental High-Speed Rail: Resumed

Another direction, rather than moving up to mag-lev, is what is called incremental high-speed rail, which was introduced earlier. Incremental high-speed rail is characterized by a right-of-way shared between passenger and freight operations, and would operate at speeds in the range of 125–150 mph. There is interest in this idea in the United States. It is largely promoted on financial grounds. The costs associated with incremental systems are substantially less, since existing rights-of-way are upgraded rather than building new rights-of-way at quite substantial costs.

In Daniel Roth's work, mentioned earlier, he recognized that achieving the cooperation that would be necessary between the freight-carrying railroads—who own the right-of-way in the United States—and public authorities, Amtrak typically, who would operate the passenger trains—was problematic.

Operations Issues for Incremental High-Speed Rail

There are many issues that one has to deal with in operating freight and high-speed trains on the same set of tracks. Passenger trains would travel 125 mph or more; freight trains usually travel much slower, so interference is an issue. Optimal track alignment and track structure would be different for the two services. Clearly, we would align and structure the track differently for a 125-mph passenger train than for a coal train with high wheel-loadings. The sophistication of signaling systems that one would need would be different.

The protection that the system would require for grade crossings would be very different. This is not so much to protect the automobiles and the trucks; the issue is the protection of the train. Why? Because an accident between, say, a freight train with a crew of four people with a truck is a very different matter than an accident between a truck and a passenger train carrying 400 passengers, given the relative values of life and freight. So we require much better safety.

Another example—let's suppose we have a hazardous material being carried by a freight train, followed by a passenger train. While the probability of a collision is very low, the negative impacts of a disaster are much higher when passengers are involved.

In addition, there are issues involving line capacity. In the last decade in the United States, freight rail demand has grown, and some capacity constraints on the links are appearing—on the right-of-way. If one adds a larger number of high-speed passenger trains to the mix, the issues of capacity become more critical. Traffic interference between high-speed passenger trains and relatively slow-speed freight trains becomes an issue. The freight railroads are fundamentally in the business of running a freight operation. If their capacity is being limited by passenger operations, the freight railroads have real economic reasons to object to additional passenger trains. Figure 30.5 highlights some of the technical and operating issues in sharing right-of-way between passenger trains and freight trains.

There is another interesting perspective on this. One would think that incremental high-speed rail, as defined here, would be the darling of the environmental community, who would say, "This is a really good idea. We get people out of their automobiles going between, say, New York and Boston or, say, between Albany and Buffalo; they will be on the train and not in the highly polluting motor vehicle." But, there is another side of that coin. What is the downside from an environmental perspective here?

FIGURE 30.5 *Sharing right-of-way between passenger and freight trains (Source: Roth, D., "Incremental High Speed Rail in the U.S.: Economic and Institutional Issues," Thesis for Master of Science in Transportation, Department of Civil and Environmental Engineering, MIT, July 1994).*

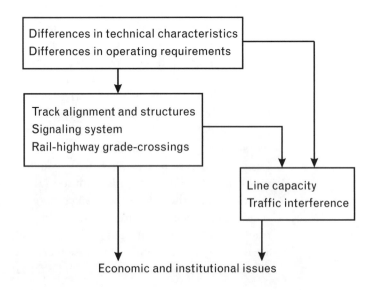

It might push freight onto trucks if the railroads did not have adequate capacity.

Exactly. If the level-of-service of the freight railroads deteriorated as a result of having the additional passenger travel, and shippers said, "Service has really deteriorated. I'm going to start using trucks," that is not an environmental gain. As always, one has to understand the systemic effects.

Amtrak Funding and Structure

Amtrak is the national provider of intercity passenger rail service in the United States, a quasi-public organization, funded primarily by the federal government. Amtrak owns no right-of-way (except the Northeast Corridor). Rather, it operates over the right-of-way owned and maintained by freight-carrying railroads. Amtrak, however, does own, maintain, and operate the equipment. Rail passengers pay through the farebox to use Amtrak services. The various states receive federal funds that are used to subsidize Amtrak's operation in addition to funds the states provide directly.

The railroads have control of train dispatching. Amtrak owns, operates, and maintains the rolling stock; the railroads own, dispatch and maintain the infrastructure. In addition, Amtrak pays the freight-carrying railroads for the use of the right-of-way. Amtrak has recently been carrying some cargo—in addition to their traditional niche in mail and small packages—greatly annoying the freight railroads.

So basically, we have here a subsidized rail passenger system. Amtrak is funded out of the farebox by users. Federal funds flow through the states who, in turn, subsidize Amtrak directly for operations. At the same time, some of the states may be investing in the rights-of-way of railroads in their states to allow passenger operations to continue.

Now, what Daniel Roth did was to look at a number of other institutional arrangements that might exist among passengers, the states, the federal government, Amtrak, and railroads, regarding equipment and ownership of right-of-way. For example, Amtrak does own the Northeast Corridor and dispatches trains. In the 1970s, Amtrak, with, of course, government support, paid for and now owns the right-of-way that exists between Boston and Washington, so in that particular example, the right-of-way, as well as equipment, is owned by Amtrak.

Perspectives of Freight Railroads on Passenger Service

There are differences of opinion among the freight-carrying railroads about incremental high-speed rail and their right-of-way. There are some who are adamantly against it. There are some who think it is a viable idea and potentially an important new market. There are a number of important issues that differentiate the railroads in this regard.

Capacity

First, we have talked about capacity and level-of-service being two important issues. If you don't have enough right-of-way capacity for your growing freight business, you are not very excited about providing any of that capacity for passenger service. So capacity is a key issue, and level-of-service, both for passengers and for freight, is a key issue. Remember that as systems operate closer and closer to capacity, the level-of-service provided is poorer and poorer—the "hockey stick" (see Figure 30.6).

Liability

Second, there is the liability concern of various railroads. What happens if there is an accident? There was a major accident in 1987 in Chase, Maryland, outside of Baltimore; an Amtrak train and a Conrail train collided, with loss of life. A Conrail engineer was under the influence at the time. Conrail thought there was going to be limited liability. However, the federal courts found them grossly negligent, and all the carefully negotiated statutes about limited liability went out the window. So the risk profile changes when a railroad has extensive passenger operations on its infrastructure.

FIGURE 30.6 *LOS degrades as volume approaches capacity.*

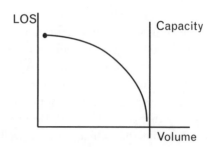

Cost-Sharing and Cost-Allocation

Third, there is the question of cost-sharing and cost-allocation. Right-of-way is being shared by freight and passenger; how do you decide who pays what for the use of that right-of-way? One could imagine, depending upon your point of view, coming up with all sorts of methods for how to do cost allocation.

At one extreme, Amtrak, the passenger carrier, argues, "This track is here already. The amounts of money you have to spend to make it useful for our service is small. You should not be charging us very much money at all." The freight railroads are saying, "Oh, you are wrong. We run our heavy trains at 60 mph. We maintain, we align, and super-elevate for those trains. The fact that we need to accommodate your high-speed trains changes everything. It's very expensive for us to provide for infrastructure for this passenger service."

So, there is a lot of negotiation on the question of cost-allocation; the economists argue about it and the engineers argue about it. How much does it really cost when a train going 120 mph goes over your track, as compared with how much it really costs when a freight train going 60 mph goes over that track?

Cost allocation is an important question in all modes. How do we allocate airport operating costs between general aviation and commercial aviation, or highway costs between automobiles and trucks? There is a whole body of literature that has developed on these questions.

Growth Path for Incremental High-Speed Rail

Another question deals with the growth path for incremental high-speed rail. There are two schools of thought on this. One school of thought says incremental high-speed rail is simply a way-station on the long-range path to dedicated true high-speed rail in this country. We are going to spend a reasonable amount of money to build incremental high-speed rail in some corridors. It will demonstrate a market for these services; it will demonstrate that railroads can handle this kind of an operation; it is going to lead, through the political process and passengers voting with their feet by riding this service, to the development, in several decades, of dedicated true high-speed rail systems around the nation. That is one school of thought.

The other school of thought says that incremental high-speed rail is as far as we ever go in the United States. Incremental high-speed rail is an end state, not a transition state. We are going to invest in incremental high-speed rail and we will have 125-, 135-, 140-mph service in the Boston-Washington corridor and a few other places, and that will be the end of that. It will be a viable service, with reasonable ridership with a reasonable subsidy, but there will never be the need for a full-blown dedicated rail system with its extraordinary cost required by a dedicated right-of-way. Incremental high-speed rail will take the political pressure off—rather than developing a constituency for dedicated high-speed rail.

Now, only time will tell which of those two points of view is correct. Some researchers are trying to look for analogs in both the transportation world and in other technologies to try to predict whether incremental high-speed rail is a transition state to dedicated high-speed rail or even mag-lev, or an end state for surface transportation in the United States (see Figure 30.7).

The 30 Key Points and Traveler Transportation

This concludes the section on traveler transportation. As with freight, a useful exercise is to think about the triad of technology, systems, and institutions, and the 30 Key Points of Part I, as they apply to the various modes and to intermodal connections among them.

FIGURE 30.7 *HSR, incremental HSR, and mag-lev.*

HSR
Dedicated service
very high speed
150–200 m.p.h.
(Europe and Japan)

Less technology   More technology

· Incremental HSR
· Shared ROW (with freight)
· Speed: 125–150 m.p.h.
· Safety issue (grade-crossing)
· Capacity issues

· MAG LEV
· Dedicated ROW
· Speed: 300 m.p.h.
· "High tech"
· Very expensive (comparatively)
· As yet unproven technology
  (commercially)

REFERENCES

1. Roth, D., "Incremental High Speed Rail in the U.S.: Economic and Institutional Issues," Thesis for Master of Science in Transportation, Department of Civil and Environmental Engineering, MIT, July 1994.

2. Thornton, R., "Beyond Planes, Trains and Automobiles: Why the United States Needs a Mag-Lev System," *Technology Review*, April 1991.

Afterword

As this is being written, the world of transportation is on the cusp of its entry into the 21st century. The intertwining of transportation and the development of human society is a long and rich story. The Internet and communications revolution notwithstanding, I am confident that transportation will continue to play a major role in economic development and the quality of life of the human species.

New technologies will doubtless impact the field and new systems concepts will overtake our current transport deployments. New institutional relationships governing and shaping the impact of transportation on the modern world will doubtless develop.

It is difficult to foresee the future of this field. Some modes will flourish and others decay, in different ways, in different parts of the world. Certainly advanced technologies and changing human needs will affect how transportation relates to modern society.

Given that, what I have tried to do in this book is to create a framework for understanding transportation systems at a basic level, so that students, faculty members, and practitioners can go forward to design the transportation systems and programs of the future, based on strong fundamentals and a good understanding of how the contemporary system works and the issues surrounding it.

At the same time, we introduce transportation as prototypical of CLIOS (complex, large, integrated, open systems), as discussed in Chapter 1, and part of the evolution toward the study of the broader field of engineering systems—focused on technology, systems, and institutions. In this direction lies the future of the study of transportation.

BIBLIOGRAPHY

1. General Transportation and Modeling

Ballou, R. H., *Business Logistics Management*, 3rd ed., Englewood Cliffs, NJ: Prentice-Hall, 1992.

Daganzo, C. F., *Fundamentals of Transportation and Traffic Operations*, Pergamon, 1997.

Homburger, W. S., *Transportation and Traffic Engineering Handbook,* 2nd ed., Institute of Transportation Engineers (ITE), Washington, D.C., 1982.

Law, A. M. and W. D. Kelton, *Simulation Modeling and Analysis*, 2nd ed., New York: McGraw-Hill, 1991.

Lieb, R., *Transportation*, 4th ed., Dome Publications, 1994.

Long, S. G., *The Annals of the American Academy of Political and Social Science: Transport at the Millennium*, Special Edition, September 1997.

Manheim, M. L., *Fundamentals of Transportation System Analysis*, Vol. 1: Basic Concepts, Cambridge, MA: The MIT Press, 1979.

Meyer, M. D. and E. J. Miller, *Urban Transportation Planning: A Decision-Oriented Approach*, New York: McGraw-Hill, Inc., 1984.

Morlok, E. K., *Introduction to Transportation Engineering and Planning*, New York: McGraw-Hill, 1978.

Sheffi, Y., *Urban Transportation Networks: Equilibrium Analysis with Mathematical Programming Methods*, Englewood Cliffs, NJ: Prentice-Hall, 1985.

2. Transportation Demand and Economics

Ben-Akiva, M., and S. Lerman, *Discrete Choice Analysis: Theory and Application to Travel Demand*, Cambridge, MA: The MIT Press, 1985.

Gómez-Ibáñez, J., W. Tye, and C. Winston, *Essays in Transportation Economics and Policy*, Brookings Institution Press, 1999.

Krugman, P., *Geography and Trade*, Cambridge, MA: The MIT Press, 1991.

Meyer, J. R., et al., *Competition in the Transportation Industries*, Cambridge, MA: Harvard University Press, 1964.

3. Transportation Statistics

Pisarski, A. E., "Commuting in America II: The Second National Report on Commuting Patterns and Trends," ENO Transportation Foundation, 1996.

"Transportation Statistics Annual Report 1997," Bureau of Transportation Statistics, U.S. Department of Transportation.

4. Highways/Automotive

Downs, A., *Stuck in Traffic—Coping with Peak-Hour Traffic Congestion*, Brookings Institution Press, 1992.

Dunn, J. A., Jr., *Driving Forces: The Automobile, Its Enemies and the Politics of Mobility*, Brookings Institution Press, 1998.

Garber, N. and L. A. Hoel, *Traffic and Highway Engineering*, St. Paul, MN: West Publishing Co., 1988.

Jones, D., D. Roos and J. Womack, *The Machine that Changed the World*, New York: Rawson Associates, 1990.

Mannering, F. and W. Kilaresky, *Principles of Highway Engineering and Traffic Analysis*, 2nd ed., New York: John Wiley and Sons, 1998.

"Curbing Gridlock: Peak Period Fees to Relieve Traffic Congestion," Vol. 1 and 2, TRB Special Report 242, Washington, D.C.: National Academy Press, 1994.

5. Intelligent Transportation Systems (ITS)

Ashok, K., "A Framework for Dynamic Traffic Prediction," Ph.D. Thesis, Massachusetts Institute of Technology, August 1996.

Branscomb, L. M. and J. H. Keller (eds.), *Converging Infrastructures: Intelligent Transportation and the National Information Infrastructure*, Cambridge, MA: The MIT Press, 1996.

Chen, K. and J. C. Miles (eds.), *ITS Handbook 2000: Recommendations from the World Road Association (PIARC)*, Norwood, MA: Artech House, 1999.

Klein, H., "Institutions, Innovations, and the Information Infrastructure: The Social Construction of Intelligent Transportation Systems in the U.S., Europe, and Japan," Ph.D. Thesis, Massachusetts Institute of Technology, June 1996.

McQueen, B. and J. McQueen, *Intelligent Transportation Systems Architectures*, Norwood, MA: Artech House, 1999.

Walker, J., ed., *Advances in Mobile Information Systems*, Norwood, MA: Artech House, 1999.

"A Strategic Plan for Intelligent Vehicle Highway Systems in the US," ITS America, 1992.

"Technologies for Intelligent Vehicle Highways," *Technology Tutorial Series*, Vol. 2, SPIE, November 1994.

6. Public Transportation

Bernick, M. and R. Cervero, *Transit Villages in the Twenty-First Century*, New York: McGraw-Hill, 1997.

Gray, G. E. and L. Hoel (eds.), *Public Transportation: Planning, Operations and Management*, 2nd ed., Englewood Cliffs, NJ: Prentice-Hall, Inc., 1992.

Hoffman, A., "Toward a Positioning Strategy for Transit Services in Metropolitan San Juan: An Initial Typology of Public Perception of Transit Options," Thesis for Master of Science in Urban Studies and Planning, Massachusetts Institute of Technology, February 1996.

Vuchic, V. R., *Urban Public Transportation Systems and Technology*, Englewood Cliffs, NJ: Prentice-Hall, 1981.

APTA 1999: Transit Fact Book, American Public Transit Association, 1999.

7. Urban Aspects: Transportation, Form, and Issues

Brookings Review, Special Issue on "The New Metropolitan Agenda," Brookings Institution Press, Fall 1998.

Downs, A., *New Visions for Metropolitan America*, Brookings Institution Press, 1994.

Garreau, J., *Edge City: Life on the New Frontier*, Doubleday, 1991.

Gybczynski, W., *City Life*, Touchstone, Simon & Schuster, 1995.

Hanson, S. (ed.), *The Geography of Urban Transportation*, 2nd ed., New York: Guilford Press, 1995.

Jacobs, J., *The Economy of Cities*, New York: Random House, 1969.

Mitchell, W., *e-topia*, Cambridge, MA: The MIT Press, 1999.

Mumford, L., *The City in History*, Harcourt Brace, 1961 (also MJF books).

Norquist, J. O., *The Wealth of Cities: Revitalizing the Centers of American Life*, Addison-Wesley, 1998.

Pucher, J. and C. Lefèvre, *The Urban Transport Crises in Europe and North America*, MacMillan Press, LTD, 1996.

Weiner, E., *Urban Transportation Planning in the U.S.: An Historical Overview*, rev. ed., U. S. DOT, 1992.

Winston, C. and S. Chad, *Alternate Route: Toward Efficient Urban Transportation*, Brookings Institution Press, 1998.

Wright, C. L., *Fast Wheels, Slow Traffic: Urban Transport Choices*, Temple University Press, 1992.

8. Passenger Rail

Lynch, T., ed., *High Speed Rail in the U.S.: Super Trains for the Millennium*, Gordon and Breach Science Publishers, 1998.

Phelan, R. S., "Construction and Maintenance Concerns for High Speed Maglev Transportation Systems," MSCE Thesis, Massachusetts Institute of Technology, June 1990.

Phelan, R. S., "High Performance Maglev Guideway Design," Ph.D. Thesis, Massachusetts Institute of Technology, January 1993.

Roth, D., "Incremental High Speed Rail in the U.S.: Economic and Institutional Issues," MST Thesis, Massachusetts Institute of Technology, July 1994.

Vranick, J., *Supertrains: Solutions to America's Transportation Gridlock*, St. Martins Press, 1991.

Wilner, F. N., *The Amtrak Story*, Omaha, NE: Simmons-Boardman Books, Inc., August 1994.

"In Pursuit of Speed: New Options for Intercity Passenger Transport," Special Report 233, Transportation Research Board, National Research Council, Washington, D.C., 1991.

9. Airports

de Neufville, R., *Airport System Planning: A Critical Look at the Methods and Experience*, The MacMillan Press, 1976.

"Airport System Capacity — Strategic Choices," TRB Special Report 226, Transportation Research Board, National Research Council, Washington, D.C., 1990.

10. Freight: Railroads and Trucking

Armstrong, J. H., *The Railroad: What It Is, What It Does*, 3rd ed., Omaha, NE: Simmons-Boardman Books, Inc., 1993.

Caplice, C. G., "An Optimization-Based Bidding Process: A New Framework for Shipper-Carrier Relationships," Ph.D. Thesis, Massachusetts Institute of Technology, June 1996.

Dong, Y., "Modeling Rail Freight Operations under Different Operating Strategies," Ph.D. Thesis, Massachusetts Institute of Technology, September 1997.

Kwon, O. K., "Managing Heterogeneous Traffic on Rail Freight Networks Incorporating the Logistics Needs of Market Segments," Ph.D. Thesis, Massachusetts Institute of Technology, August 1994.

Railroad Facts, 1999 Edition, Association of American Railroads, Washington, D.C., October 1999.

11. Maritime Freight

Stopford, M., *Maritime Economics*, 2nd ed., New York: Routledge, 1997.

"Intermodal Marine Container Transportation Impediments and Opportunities," TRB Special Report 236, National Academy Press, Washington, D.C., 1992.

12. Intermodal Freight

Muller, G., *Intermodal Freight Transportation*, 4th ed., ENO Transportation Foundation, 1997.

13. Regional Transportation Issues

Yaro, R. D. and T. Hiss, "A Region at Risk: The Third Regional Plan for the New York-New Jersey-Connecticut Metropolitan Area," Regional Plan Association, Washington, D.C.: Island Press, 1996.

"New England Transportation Initiative: Final Report," Cambridge Systematics, Inc., February 1995.

"Turning Point: Special Report on 'The Boston Conference: Shaping the Accessible Region'," Special section of *The Boston Globe*, October 10, 1994.

14. Transportation in Developing Countries

Dimitrou, H., *Urban Transport Planning: A Developmental Approach*, New York: Routledge, 1992.

Dutt, P., "A Standards-Based Methodology for Urban Transportation Planning in Developing Countries," Ph.D. Thesis, Massachusetts Institute of Technology, September 1995.

15. Sustainable Transportation/Energy

Greene, D. L., *Transportation and Energy*, ENO Transportation Foundation, 1996.

Sperling, D., *Future Drive: Electric Vehicles and Sustainable Transportation*, Washington, D.C., Island Press, 1995.

Sperling, D. and S. A. Shaheen (eds.), *Transportation and Energy: Strategies for a Sustainable Transportation System*, ACEEE, Washington, D.C., 1995.

Transportation Research Board/National Research Council, "Expanding Metropolitan Highways: Implications for Air Quality and Energy Use," Special Report 245, Washington, D.C.: National Academy Press, 1995.

Transportation Research Board/National Research Council, "Toward a Sustainable Future: Addressing the Long-Term Effects of Motor Vehicle Transportation on Climate and Ecology," Special Report 251, Washington, D.C.: National Academy Press, 1997.

16. Transportation Technology

Office of Economic and Policy Analysis, The Port Authority of New York and New Jersey, *The Technology Review Study: Significant Emerging Technologies and Their Impacts on the Port Authority*, October 1994.

Scientific American, "Special Issue: The Future of Transportation," October 1997.

"Technology/Research and Development Forum on Future Directions in Transportation R&D," Washington, D.C.: National Academy Press, 1995.

"Transportation Science and Technology Strategy," Committee on Transportation Research and Development, Intermodal Transportation Science and Technology Strategy Team, National Science and Technology Council, September 1997.

"U. S. Department of Transportation Research and Development Plan," 1st ed., May 1999.

17. Transportation Organizations/Governance/Institutions

Luberoff, D. and A. Altshuler, *Mega-Project—A Political History of Boston's Multibillion Dollar Artery/Tunnel Project*, Taubman Center for State and Local Government, Harvard University, Cambridge, MA, 1996.

National Transportation Organizations: Their Role in the Policy Development and Implementation Process, ENO Transportation Foundation, 1997.

18. Transportation History

Goddard, S., *Getting There: The Epic Struggle between Road and Rail in the American Century*, Chicago: The University of Chicago Press, 1994.

Lay, M. G., *Ways of the World: A History of the World's Roads and of the Vehicles that Used Them*, New Brunswick, NJ: Rutgers University Press, 1992.

Sobel, D., *Longitude*, New York: Walker & Co., 1995.

Transportation History and TRB's 75th Anniversary, containing Hoel, L. A., "Historical Overview of U.S. Passenger Transportation," and Sussman, J. M., "Transportation's Rich History and Challenging Future — Moving Goods," Transportation Research Circular, Number 461, Transportation Research Board/National Research Council, Washington, D.C., August 1996.

America's Highways: 1776–1976, U. S. Department of Transportation, Federal Highway Administration, Washington, D.C., 1976.

Vance, J. E., Jr., *Capturing the Horizon: The Historical Geography of Transportation since the Sixteenth Century*, John Hagelin Press, 1995.

19. Management/Economic Trends and Ideas

Chandler, A. D., Jr., *Strategy and Structure: Chapters in the History of the Industrial Enterprise*, Cambridge, MA: The MIT Press, 1962.

Chandler, A. D., Jr., *The Visible Hand: The Managerial Revolution in American Business*, Cambridge, MA: Harvard University Press, 1977.

Day, G. S. and D. J. Reibstein (eds.), *Wharton on Dynamic Competitive Strategy*, New York: John Wiley & Sons, 1997.

Drucker, P., *Management: Tasks, Responsibilities, Practices*, New York: Harper & Row, 1974.

Hamel, G. and C. K. Prahalad, *Competing for the Future*, Boston, MA: Harvard Business School Press, 1994.

Hammer, M. and J. Champy, *Re-engineering the Corporation: A Manifesto for Business Revolution*, Harper Collins Publishers, Inc., 1993.

Hardy, C., *The Age of Unreason*, Boston, MA: Harvard Business School Press, 1990.

Hughes, T., *Rescuing Prometheus*, New York: Pantheon Books, 1998.

Kanter, R. M., *World Class: Thriving Locally in the Global Economy*, New York: Simon & Schuster, 1995.

Levitt, T., *Thinking about Management*, The Free Press, 1991.

Micklethwait, J. and A. Wooldridge, *The Witch Doctors: Making Sense of the Management Gurus*, Times Books, Random House, 1996.

Mitchell, W., *City of Bits: Space, Place and the Infobahn*, Cambridge, MA: The MIT Press, 1995.

Negroponte, N., *Being Digital*, Knopf, 1995.

Osborne, P., and T. Gaebler, *Reinventing Government—How Government Can Get More Effective and Efficient, in Partnership with the Private Sector*, Addison Wesley Publishing Co., Inc., 1992.

Perrow, C., *Normal Accidents: Living with High-Risk Technologies*, Basic Books, 1984.

Porter, M., *Competitive Strategy: Techniques for Analyzing Industries and Competitors*, The Free Press, 1980.

Schon, D. A., *The Reflective Practitioner: How Professionals Think in Action*, Basic Books, 1983.

Senge, P., *The Fifth Discipline: The Art and Practice of the Learning Organization*, New York: Currency Doubleday, 1990.

Swartz, P., *The Art of the Long View: Planning for the Future in an Uncertain World*, New York: Currency Doubleday, 1990.

About the Author

Dr. Joseph M. Sussman is the JR East Professor (endowed by the East Japan Railway Company) in the department of civil and environmental engineering and the engineering systems division at the Massachusetts Institute of Technology, where he has served as a faculty member for 32 years.

Dr. Sussman specializes in planning, investment analysis, operations, management, design, and maintenance of large-scale transportation systems, working in many modal environments. His research in rail service reliability, rail operations, freight car maintenance and track maintenance, high-speed rail, and rail risk assessment in the United States and Japan has had a major impact on the railroad industry and has resulted in several prize-winning papers.

He has worked on intelligent transportation systems (ITS), helping to build the United States' national program. While serving as the first Distinguished University Scholar at IVHS America (1991–1992), he was a member of the core group that wrote the strategic plan for IVHS in the United States, a 20-year plan for research, development, testing, and deployment that has shaped the United States ITS program. He has worked on the development of an "intelligent corridor" in Bangkok; on the relationship of ITS to the National Information Infrastructure (NII); on a comparative analysis of ITS programs in western Europe, Japan, and the United States; on a strategic plan for ITS/commercial vehicle deployment in Maryland; and on methods for building regional ITS architectures. He currently writes a column entitled "Thoughts on ITS" for the *ITS Quarterly* and served as the program chairman for the ITS America Annual Meeting in May 2000 in Boston.

Dr. Sussman earned a B.C.E. from City College of New York in 1961, an M.S.C.E. from the University of New Hampshire in 1963, and a Ph.D. in civil engineering systems from MIT in 1967. He joined the MIT faculty in 1967. From 1977 to 1979, he served as the Associate Dean of Engineering for Educational Programs. He served as head of the Department of Civil Engineering from 1980 to 1985, and as the director of the Center for Transportation Studies (CTS) from 1986 to 1991.

Dr. Sussman is a member of the American Society of Civil Engineers, Transportation Research Forum, Transportation Research Board (TRB Executive Committee, 1991–1998), and ITS America (Board of Directors, 1995–present). He has authored numerous publications and has lectured and consulted extensively with transportation companies, government agencies, and commissions in the United States and abroad. He is a founder of Multisystems in Cambridge, Massachusetts. He is also the founding and current chair of the TRB review committee for the federal transportation R&D strategy.

Index

Recent Titles in the Artech House ITS Library

John Walker, Series Editor

Advances in Mobile Information Systems, John Walker, editor

Incident Management in Intelligent Transportation Systems,
Kaan Ozbay and Pushkin Kachroo

Intelligent Transportation Systems Architectures, Bob McQueen and
Judy McQueen

Introduction to Transportation Systems, Joseph Sussman

*ITS Handbook 2000: Recommendations from the World Road Association
(PIARC),* PIARC Committee on Intelligent Transport (Edited by Kan Chen and
John C. Miles)

Positioning Systems in Intelligent Transportation Systems,
Chris Drane and Chris Rizos

Smart Highways, Smart Cars, Richard Whelan

Tomorrow's Transportation: Changing Cities, Economies, and Lives,
William L. Garrison and Jerry D. Ward

Vehicle Location and Navigation Systems, Yilin Zhao

Wireless Communications for Intelligent Transportation Systems,
Scott D. Elliott and Daniel J. Dailey

For further information on these and other Artech House titles, including
 previously considered out-of-print books now available through our
In-Print-Forever® (IPF®) program, contact:

Artech House
685 Canton Street
Norwood, MA 02062
Phone: 781-769-9750
Fax: 781-769-6334
e-mail: artech@artechhouse.com

Artech House
46 Gillingham Street
London SW1V 1AH UK
Phone: +44 (0)20 7596-8750
Fax: +44 (0)20 7630 0166
e-mail: artech-uk@artechhouse.com

Find us on the World Wide Web at:
www.artechhouse.com